Life in the
BALANCE

Humanity

and the

Biodiversity Crisis

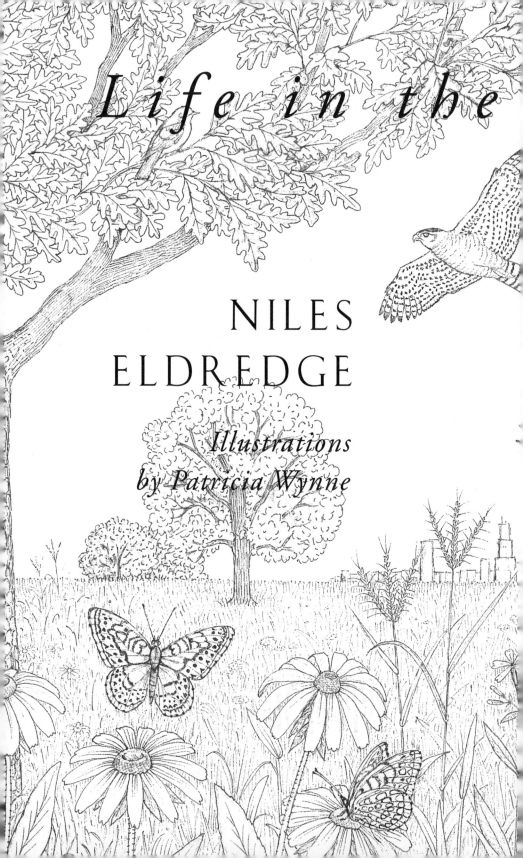

Life in the

NILES
ELDREDGE

*Illustrations
by Patricia Wynne*

BALANCE

Humanity and the Biodiversity Crisis

A PETER N. NEVRAUMONT BOOK

PRINCETON UNIVERSITY PRESS

Princeton, New Jersey

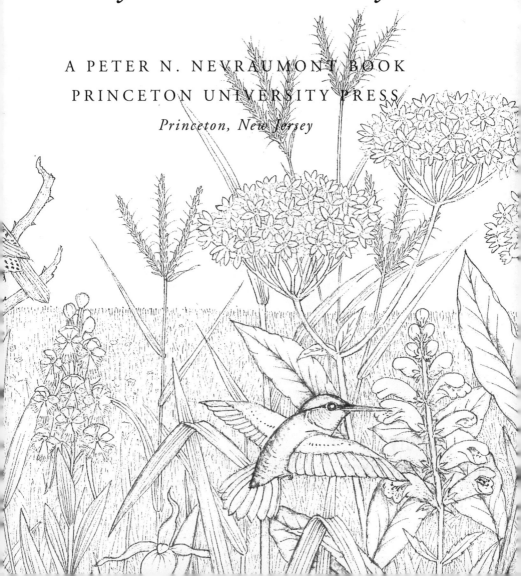

Published by Princeton University Press
41 William Street
Princeton, New Jersey 08540

Second printing, and first paperback printing, 2000

Paperback ISBN 0-691-05009-0

The Library of Congress has cataloged the cloth edition of this book as follows

Eldredge, Niles.
Life in the balance : humanity and the biodiversity crisis / Niles
Eldredge : illustrations by Patricia Wynne.
p. cm.
"A Peter N. Nevraumont book."
Includes bibliographical references and index.
ISBN 0-691-00125-1 (cloth : alk. paper)
1. Biological diversity conservation. 2. Human ecology.
I. Title.
QH75.E58 1998   97-52087
333.95—dc21

The paper used in this publication meets the minimum requirements of
ANSI/NISO Z39.48-1992 (R1997) (*Permanence of Paper*)

http://pup.princeton.edu

Created and Produced by Nevraumont Publishing Company
New York, New York

President: Ann J. Perrini

Jacket and book design by Jaye Zimet

Printed in the United States of America

3 5 7 9 10 8 6 4 2

CONTENTS

P R E F A C E
THE BIODIVERSITY CRISIS

Life is beautiful. The living world surrounds us, and we human beings are a part of it. We depend on the living world for our food and medicines, the oxygen that we breathe, and the elements that form our bodies. Earth is our home, and all Earth's species are our family, for we have evolved along with every other living thing.

Yet all is not well with life on Earth. We are losing roughly 30,000 species of plants and animals a year as ecosystems are disturbed and individual species are overhunted or overharvested. We need to ask ourselves, What does the living world mean to us? Why should we care if ecosystems degrade and species are lost? What is causing species extinction? And what can we do to stem the tide of this latest crisis in the living world?

WHAT IS BIODIVERSITY?
Life is abundant, colorful, and richly diverse. "Biodiversity" is that rich spectrum of life—all the world's species ranging from the smallest bacteria to the giant redwoods; from the algae of the sea to the wild dogs of the African savannas; from the worms of the soil to the falcons soaring overhead. Biodiversity embraces all the bacteria and other microbes, many of which perform vital chemical functions to keep ecosystems functioning. Biodiversity also includes the green plants that produce oxygen through photosynthesis, trapping solar energy and storing it in the form of sugars that are the base of energy resources for all other forms of life.

Biodiversity includes the fungi—mushrooms and their kin, responsible for decay, nutrient recycling, and chemical building blocks so vital to keep life going. And biodiversity also encompasses the animals—everything from sponges to birds and mammals, including our own species, *Homo sapiens*, and our closest living evolutionary relatives, the chimps, gorillas, and orangutans. No one knows for sure how many species form this evolutionary spectrum of life. Scientists from institutions like the American Museum of Natural History have counted at least 1.75 million species so far, but we know there must be at least 10 million living species, and some scientists think there are a great deal more.

Species are the players in the great game of life that takes place in all the world's ecosystems. Each ecosystem—every lake, bog, and stream; every alpine meadow; every patch of prairie and stand of forest—is home to many different species, each of which plays a specific role as its own ecological niche. Energy flows through the system as some animals consume the plants that have trapped solar energy through photosynthesis; these animals are eaten by other animals; and eventually all are destined to die and be recycled by the microbes and fungi. This is the ecological aspect of biodiversity—the vast spectrum of ecological systems ranging from the polar ice caps with relatively few species to the forests and grasslands of the Tropics, which teem with species. Life clings to the highest mountain tops and extends down to the deepest trenches of the ocean.

WHY SHOULD WE CARE ABOUT BIODIVERSITY?
Ten thousand years ago, humans began to cultivate crops, drastically changing the human ecological niche. Before agriculture, people lived as hunter-gatherers, completely dependent on the foods and other resources produced by the ecosystems in which they lived. With the invention of agriculture, human beings no longer lived inside local ecosystems.

Yet humanity still depends heavily on Earth's living species and ecosystems. People around the world utilize over 40,000 species every day—and most of these are plants. [See Appendix II] All the crops that we cultivate—corn, wheat, potatoes, tomatoes, apples, pears, oranges and on and on—are domesticated from wild species. Crop geneticists use wild populations to replenish genetic variation of domesticated species, to improve crop yields, and to increase both resistance to disease and the ability to grow in many different climates.

Wild species—including microbes, plants, and even animals—are also crucial to the search for medicines. Though it is possible in many cases to synthesize drugs in the laboratory, we must first know about the existence of a chemical useful in fighting disease before we can make it ourselves. Nature is a natural pharmacopoeia, and new drugs and medicines are being discovered in the wild all the time. Aspirin comes from the bark of willow trees. Penicillin comes from a mold, a type of fungus. Newer drugs, such as the cancer-fighting substances found in the fruit of the African sausage tree, the bark of the Pacific yew tree, or the Madagascar periwinkle—a simple wildflower—show how dependent we continue to be on wild species for the quality of human health.

Although we humans no longer live inside local ecosystems, we still need them to provide the essential "ecosystem services," on which all living creatures—including ourselves—depend. The production of oxygen, the cycling of freshwaters, the prevention of soil erosion, the fixation of nitrogen are all vital functions performed inside healthy local ecosystems. In the dry tropical savannas, the only creatures that can perform the vital task of decomposition and decay are the bacteria and fungi living either inside termites or in their nests. Without the recycling of carbon and many other elements, life—including human life on Earth—would quickly come to an end.

We humans also value life around us—beautiful, eye-catching species, gorgeous intact wild places—for its intrinsic worth. Something within us recognizes that we are connected to this natural world and that we gain peace and pleasure from being in it whenever we can. For these reasons alone, we must care about what is happening to life on this planet.

THREATS TO BIODIVERSITY Life has suffered

at least five major mass extinctions in the past. The most recent—and most famous—event occurred about 65 million years ago, when the last of the dinosaurs and many other species on land and in the sea became extinct. Most scientists agree that a collision between a meteor or comet and Earth caused that extinction episode. After the dust finally settled, life eventually sprang back, with new species evolving to take the place of those driven to extinction. Mammals, which had played only modest roles in ecosystems dominated by dinosaurs, now took center stage in terrestrial ecosystems.

Evolution will replace extinct species only after whatever causes a burst of extinction disappears. Today's biodiversity crisis—the "Sixth Extinction" threatening so many of the world's species—is being caused, not by meteors or other environmental changes, but by ourselves, our species *Homo sapiens*. Only some time long after we change our behavior (or become extinct ourselves) will evolution eventually replace the species and rebuild the ecosystems that have already been lost or damaged so severely.

Ten thousand years ago, when human beings first started using agriculture and stopped living in local ecosystems, there were only about 5 million people on the planet. Today there are nearly 6 billion. Together with the unequal distribution of wealth and resources, the tremendous growth in human population is the root cause of the biodiversity crisis, the Sixth Extinction.

We humans have transformed the very face of the planet. Agriculture has triggered the population explosion, enabling the rise of civilizations and the growth of towns and cities. Clearing forests and grasslands for agriculture and the spread of those towns and cities has meant the end for many ecosystems—and species—the world over. We continue to clear the land for more agriculture to feed more people. We also continue to harvest timber for building supplies, paper manufacture, and, especially in poorer nations, simply for firewood. We are destroying thousands of acres of forest a year, not just in the tropics, but also right here in the United States.

We are also overharvesting the world's oceans. Most of the best fisheries are now severely depleted, threatening not only the livelihoods of the world's fishermen, but also the future of these oceanic and freshwater ecosystems and any hope we may have of continuing to rely on these precious food resources in the future.

As human beings have spread around the globe, we have taken other species with us, deliberately transporting domesticated animals and plants as well as accidentally introducing a number of disease-causing microbes and other species to foreign ecosystems. Some of these alien species cause no harm. Others—such as the European zebra mussel, which clogs pipes and replaces native species in North America—are not so benign. The brown tree snake of the western Pacific was introduced to the island of Guam during World War II, where it has already driven several species of native birds to extinction. Alien species often cause great ecological disturbance and are a side effect of the growth and spread of human populations around the globe.

These activities of a human society based on agriculture and high technology have accounted for the growing loss of species and the mounting numbers of disturbed ecosystems. Hundreds of species are known to have become extinct, and many thousands are suspected of having disappeared, as remote ecosystems have been destroyed before scientists have had a chance to study what was living there. People are not purely the villains of the biodiversity crisis: Many cultures—hunter-gatherers such as the BA-Aka (Pygmies) and the San (Bushmen) of Africa and the Yanomami of South America; as well as pastoralists, nomads, and pre-high-technological cultures in all the habitable regions of the world—have witnessed the loss of their traditions. Even when the peoples themselves survive, their cultures are lost to extinction.

WHAT CAN WE DO? There is hope for the future, even if the toll of destruction and loss of biodiversity already seems high. We human beings have accomplished much that is positive through the invention of agriculture. Our scientific name, *Homo sapiens*, means "wise humans." We now must understand what has happened to life on Earth as a side effect of our success; and we must decide what we can do about it. Once again, we must use our brains.

We need the concept of "enough." There are already too many humans on the planet for everyone to be able to live like a middle-class American. Stabilization of the human population—primarily through education, economic development, and the economic empowerment of women—is the way to achieve that goal. There are hopeful signs that human population growth has already begun to slow sooner than had been predicted.

We need to encourage the economic development of poorer nations, but all economic development in all nations must be *sustainable:* We must place realistic values on all resources—including ecosystems and species—that we use. We must take care not to overharvest, to replace and replant as we take from the living world, just as if we were still living inside local ecosystems.

We must continue to set aside wild places to conserve ecosystems and species, but in doing so, we must always take into account the economic lives of local peoples—lumbermen in the Pacific Northwest, farming villagers in Southeast Asia, *all* people. Conservation efforts work only with the support of local peoples, who must see benefits for themselves in leaving forests untouched and animals not hunted. We can learn again to live in harmony and equilibrium with Earth and all its living creatures. We must—our future depends on it.

THE FOUR QUESTIONS For the past several years, I have been engaged in two parallel projects, both intended to bring the four basic questions surrounding the term "biodiversity" to a wide audience. The questions are: *1.* What is biodiversity? *2.* What are its values and meaning to human life, or, why should we care about biodiversity? *3.* What precisely threatens biodiversity? And, finally, *4.* What can we do to stem the tide of the Sixth Extinction? These are the questions that together make biodiversity a pressing concern for all humanity, issues to be faced by the collective body politic as we enter the Third Millennium. That body politic needs to be informed—hence my involvement in these parallel projects.

The two projects are a new exhibition on biodiversity, opened in the spring of 1998 at the American Museum of Natural History in New York City, where I have been a curator on the scientific staff since 1969, and this book you are holding now. The exhibition, also entitled *Life in the Balance*, is the Museum's largest at over 11,000 square feet. It is also our first "issues" hall, where we depart from traditional norms of depicting Nature pristine. Instead, we reveal how human activity over the past 10,000 years has radically transformed the ecosystems of planet Earth and in doing so has caused the demise of a mounting number of species. This number threatens to crescendo into the greatest loss of species since the dinosaurs and fellow Cretaceous species were doomed by a collision between Earth and one or more extraterrestrial objects 65 million years ago.

As the chief scientist charged with developing the content for the exhibition, I have learned that exhibitions are vastly different media—modes of communication—from the lecturing and writing that have filled the greater part of my professional career. Museums are places; they are filled with objects. The best (meaning the most relevant and dazzling) are put on display to convey the message. Writing is kept to a minimum: It is notorious that labels over 50 words (some would say even fewer) are seldom read in their entirety. Instead, we rely on our specimens—supplemented by films, computer interactives, sounds, and even smells—to explore the four questions and to bring the subject of biodiversity alive.

I knew at the outset that I had to write a book on the very same four questions as our team wrestled with the often daunting task of mounting the exhibition. For the questions—and all the issues surrounding them—are so vast and complex that they demand fuller exploration with the written word. Exhibitions and books (lectures and films as well) complement each other. Although I never intended this book to act as a catalogue for the exhibition, I am, at this writing, astonished at how very different in detail, yet faithful in overall purpose of exploring the four questions, this book and the American Museum's exhibition have become.

Absolutely the only words written for both projects are those you have just read at the outset of this preface. This mini essay on the four questions was originally written as a rough draft of the script for the introductory film for the hall. The essence of those words survives in the final film product. The words themselves survive here. From there, the two projects diverge widely in their approach to the four questions.

Visitors to the Museum's exhibition will, indeed, be struck by its central organizing factor: Life comes in two distinct, albeit interrelated, modes, evolutionary and ecological. There is a spectrum of life's species, and another of life's ecosystems. The evolutionary tree of life can be read as a scorecard of the "players" in the "game of life," which is played out in the ecosystems themselves. Visitors to the museum will see an entire wall exhibiting the evolutionary array of life (a wall corresponding to chapter 3 of this book). Opposite that wall, we depict the parallel spectrum of the world's ecosystems, largely by film format, with the major exception of a gigantic walk-through 21st century diorama of a Central African Republic rain forest, replete with accurate reconstructions of plants and animals, films, and even odors. That section of the hall corresponds to chapter 4 of this book.

If the overall structure of the two projects is similar, details of the content diverge, in places radically so. Most important, to my mind, I have been at liberty to strike a personal tone in this book that was, of course, utterly impossible and inappropriate in the exhibition. No one alive has encyclopedic knowledge of all the major groups of living organisms and all the world's ecosystems. With the hall, I had many colleagues to share the daunting task of marshaling all the facts of evolutionary and ecological natural history; they included a small inner circle (Joel Cracraft, Naomi Echental, Francesca Grifo, Sidney Horenstein, Sam Taylor, Willard Whitson) plus, of course, the entire scientific staff of the American Museum of Natural History.

For this book, however, I have relied much more on my knowledge and scientific research as an invertebrate paleontologist; my experiences in the laboratory and around the world; and the scientific literature. I have decided to devote chapter 1 to a portion of the world that I have come to love during the past 6 years: Botswana, with its sere Kalahari and lush and verdant and very wet Okavango Delta. I have escorted ecotourists there and conducted research for the exhibition there as well. The "Tales from the Swamp" that directly follows this preface raises all four questions, epitomizing in microcosmic form the cosmic issues that lie at the heart of biodiversity. For good and sufficient reason, the Kalahari and Okavango stories play a minimal role in the museum's exhibition, but I love these stories as I love the places themselves, and the Kalahari-Okavango takes center position here in this book.

After the microcosmic exploration of all four questions in Botswana, I turn next to the book's longest section, What is Biodiversity? Precisely because we can look at biodiversity from the two distinctly different vantage points of evolution and ecology, I have written chapter 2 to look for the *connections* between ecology and evolution. Indeed, I find it next to impossible to think of evolution *except* in an ecological context, as my discussions of natural selection and speciation are intended to bring out.

Chapters 3 and 4, as already remarked, give a personalized tour through the spectrum of bacteria to mammals (evolutionary biodiversity, chap. 3), and then from the Arctic tundra down to the Tropics (ecological biodiversity, chap. 4). Actually, I confess that boredom with the conventional "bacteria through mammals" led me to try it the other way, from ourselves, the mammalian species *Homo sapiens* down to bacteria—a luxury likewise unavailable to me when planning the exhibition's "Wall of Life."

Keeping to the personal, I explore the next two questions, "Why should we care?" and "What is threatening biodiversity?" together in chapter 5. The two questions are deeply related, for the principal reason why we humans *don't* seem to care that much is that we have indeed been living outside of local ecosystems—effectively outside of Nature—since we invented agriculture some 10,000 years ago. Our transformation of the surface of the globe—largely for agricultural purposes, with the side effect of the build-up of large urban and suburban population centers—masks the critical importance of vital, functioning ecosystems for continued human health and well-being and at the same time is causing this tremendous yearly loss of species that is the Sixth Extinction.

Finally—and still on a heartfelt, personal level—I address the fourth question, What can we do to stem the tide of the Sixth Extinction? Some of the answers, as drafted in my call-for-action manifesto in chapter 6, may strike you as familiar and perhaps even mundane, but they are the very stuff of all concerted action—action that needs the involvement of people from all nations, from all classes, and from all walks of life.

That is why I have written this book and helped to produce the American Museum's exhibition of *Life in the Balance.* By all means, please visit the exhibition when you are in New York City. It will be there for many years. As you read this book, think of these issues, the four questions. Think about where we are going as we enter the 21st century. Think about how we humans fit into the natural world, and think about what we—as individuals, groups, and nations—can do to stem the tide of the Sixth Extinction.

Chapter 1 TALES FROM THE SWAMP PART I: THE BIODIVERSITY CRISIS IN MICROCOSM

Tucked away in the northern reaches of Botswana—a landlocked country about the size of France seated just to the north of South Africa—lies the closest thing to Eden left on the planet, the Okavango Delta. Here rich droves of wildlife still roam the grasslands, surrounded by beds of reeds and papyrus and stands of majestic gallery forest that line the myriad riverine channels braiding their way through the delta. It is a scene right out of Olduvai Gorge 2.5 million years ago, and a remnant of the stage on which our own species, Homo sapiens, *evolved in the most recent act of the human evolutionary drama, played out in Africa about 125,000 years ago.*

If ever a place deserved our attention, even our reverence, it is the Okavango Delta. The most recent extension of the famed East African Rift Valley system, where so many ancient hominid fossils have been found, the delta tells us precisely what our ancient homeland—the environment in which we evolved and, in a cultural sense, grew up—was like. To know the Okavango is to know ourselves, where we came from, how we have come to be what we are, what our relation to the natural world was, is, and most likely will be in the future.

But there is trouble in Eden. The Okavango stands imperiled by the same basic roster of threats faced by ecosystems throughout the world: overdevelopment and overuse by humans, coupled with the more usual vicissitudes of climatic change. Of these, by far the strongest threat comes from ourselves, as we poise on the brink of destroying our very birthplace. The humans responsible are not the local hunter-gathering San (Bushmen), who had until very recently been living in small bands in a rhythmic harmony with the dynamics of the local ecosystems of the delta and its drier surrounds, the scrub and grasslands of the Kalahari. Indeed, the San are as threatened with extinction—culturally as well as biologically—as are the other species of the region.

Rather, the threat comes from the pastoral peoples—initially black African Bantu-derived tribes immigrating southward, followed soon after by European colonists coming up from the Cape region of what is now South Africa. Above all else, it is modern technology—expressed in agricultural practice, water control technology, hunting weaponry, and even the planes and cars transporting tourists eager for a glimpse of the Okavango's bounties—that threatens this last vestige of Eden. It is as if our species, striking out from Africa 100,000 years ago, wending our way around the world and only recently showing up in numbers back in our birth land—now equipped with all the knowledge and technology of the modern world—has forgotten utterly how to live in the very place where we grew up.

The stories of the Okavango are our stories: Our beginnings, to be sure, but also where we are right now, what we are doing to Earth right now, and what that means for our own future. Together, the tales of the Okavango are a microcosm of the human experience, from the dim reaches of time over 2.5 million years ago, right on up past the now and across the millennial divide. I can think of no better way to begin our examination of the global issues of mounting species loss, environmental degradation, and what they mean for the human future than the specifics of the modern Botswana story. All the glories and all the perils faced by the living world appear in microcosm in these tales from the swamp, tales of life hanging in the balance.

THE PHYSICAL SETTING The Okavango Delta is a bit of a geographic oddity. Rising in the highlands of western Angola to the northwest, the Cubango River first flows south, then strikes a southeastern course. Leaving Angola, what is by now called the Okavango River traverses the Caprivi Strip, itself a modern geopolitical oddity. When the Germans controlled the colony they called German Southwest Africa (today's Namibia), they insisted on access to both the Atlantic and Indian Oceans. Namibia fronts on the Atlantic, but to reach the Indian Ocean on the other side of the southern African continent, the Germans annexed a narrow right-of-way, the Caprivi Strip, to reach the Zambezi River. The Zambezi flows out to the Indian Ocean via Mozambique, after forming the Zimbabwe-Zambia border. That the Zambezi cascades over Victoria Falls—a drop of over 90 meters—didn't seem to faze the German colonialists of the day.

It is what happens next to the Okavango that is so unusual. Entering Botswana and continuing to flow southeastward, the river flows through a 10-kilometer wide, reed and papyrus-choked flatland surrounded in places by lush grasslands, where the river's floodplain meets the surrounding drylands. Then, after some 100 kilometers, the river divides into several branches, defining the Okavango Delta proper: a fan-shaped system of waterways, grasslands, scrublands and riverine forest some 170 kilometers wide and 140 kilometers long. On all sides, there is nothing but dry scrublands and grasslands. The Okavango Delta is in reality a gigantic oasis. [Figure 1] At its bottom lies Maun, by far the most substantial settlement of the region and the jumping off point for all who would venture into the delta itself.

Say "delta," and most people think of the Mississippi, the Danube, Rhine or the Nile: Deltas form where mighty rivers meet the sea, dumping vast quantities of mud and silt eroded from the continental interiors that these rivers drain. Here, in northern Botswana, we have something very different: A river that never reaches the sea; a river that simply stops, emptying into the sands of the Kalahari. Some of its waters percolate into the

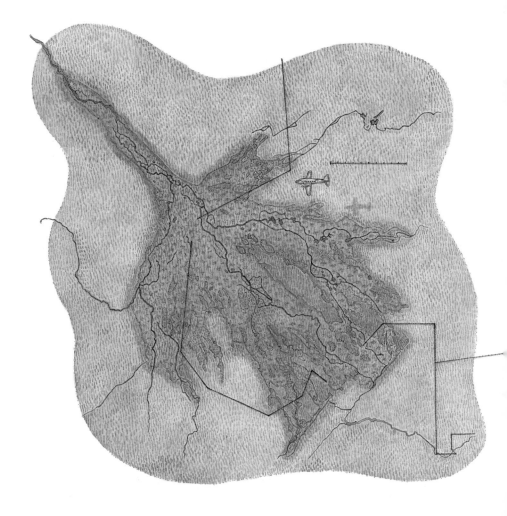

[Figure 1] *The lush Okavango Delta surrounded by the dry Kalahari. The parallel faults outlining the narrow "panhandle," as well as the boundary faults of the Delta proper (running perpendicular to the panhandle), stand out clearly on this map. A DC-3 airplane still takes ecotourists to the Delta.*

subsurface (though how much is a bone of contention). Most of it evaporates into the air that stays so dry for most of the year in this region of southern Africa. When the rains fall in Angola, swift currents reach all the way down to Maun, and the main channels—home to hippos, crocodiles, tiger fish and bream—can be up to 3 or 4 meters deep. The water is crystal clear, fresh in every sense of the word. A tenderfoot Western-world visitor can drink it straight, with little fear of contracting even the otherwise ubiquitous bacterial strains of childhood diarrhea that regularly fell adult travelers to foreign climes. The reason: The waters of the Okavango Delta are filtered through its bottom sands, and human population density is still sufficiently low in the delta that transmission of human disease pathogens is correspondingly curtailed—at least so far.

What's going on? Most of the world's other interior basins—the Great Salt Lake, the Aral, Caspian, Black, and Dead Seas—are saltier than the oceans. To stay fresh, lakes must flow. Why isn't the Okavango also salty, and why, for that matter, is it mostly drylands and wetlands bordering on fast-moving streams? The answer lies in the delicate balance between the tilt of Earth's surface, which keeps the streams running, and the dry climate, which insures that the moving waters evaporate before they get a chance to form a stagnant, and thus salty, pool.

Visitors to the delta are invariably shocked to learn that the region experiences, on average, one hefty earthquake every day. The scene is so placid, so calm, and you cannot actually feel the deep-seated tremors, cushioned as they are by the thick pile of sand and mud lying below the delta's surface. There are none of the telltale signs of active seismic (earthquake) activity. Go to California, and the fresh, jagged look to many of the hills literally screams out the constant upheaval of the landscape. Nor are there any volcanoes remotely near the delta, even though volcanoes and earthquakes typically go hand-in-hand. Nonetheless, the ground under the delta is constantly moving, sinking a bit every day. And although there are no volcanoes there now, volcanoes definitely lie in the delta's future. It may take a few million years, but the delta is destined to look like the East African Rift Valley system, with huge volcanoes like Mt. Kenya and Mt. Kilimanjaro overlooking open savannas.

The Okavango Delta, it seems, is the newest part of that thousands-of-miles long crack that starts in Ethiopia, running in from the sea and down the eastern region of the African continent. Until recently, geologists thought that the crack died out somewhere in Mozambique, but sensitive seismographs began monitoring the earthquakes of northern Botswana, and we now know that the Rift Valley system takes a sharp westward turn, running through the region of Victoria Falls and continuing to the present site of the Okavango Delta.

The African Rift Valley system is plate tectonics in action: Africa is being torn apart, as the East African crust (plate) begins to separate from the main crustal core of the African continent. As the plates move apart, the region between them drops down, forming a slowly widening chasm. The end result: Oceanic waters will flood the African interior (as they are in Ethiopia, where the crack began); what is now dry land will become an oceanic deep, and what has been for billions of years a single massive structure will become two smaller continents.

Nor is this the first such event to strike Africa. Only 160 million years ago (only!), Africa was part of an enormous single continental structure, Gondwanaland (meaning "Land of the Gonds," an indigenous Indian people living in the region where the early convincing evidence of the existence of the supercontinent was discovered). Gondwanaland once included Africa, South America, Antarctica, Madagascar, and India, conjoined into a single mega-continent. The events of the African Rift Valley system over the past 20 million years simply continue this complex process of fragmentation of the old supercontinent, and the Okavango Delta is merely the youngest manifestation of its effects.

The Okavango Delta is framed by faults: one on its northern edge, where the river divides and begins fanning out over the landscape, and the other in the south, forming the southern edge of the delta. The delta is there because the land between these faults is submerging, bit by bit, with those daily jolts. The faults have been active, and the land has been sinking, for only some 5,000 years, but already 3,000 meters of sediment lie below the surface. As the landscape sinks, the rivers bring in new supplies of sand and mud, keeping the surface level of the delta up just about to the height of the surrounding Kalahari surface—another equilibrial balance between river and landscape that helps make the Okavango Delta system so utterly unique.

Five thousand years ago the Okavango Delta didn't exist. In its place—but extending much farther south—was a real lake, an enormous expanse of inland waterway that did in fact eventually empty out to the sea. Feeding it were several rivers: the Okavango, naturally enough, but also the Chobe and Zambezi Rivers to the east of the Okavango. The lake's outlet was to the south, along the present day Limpopo River (which Rudyard Kipling called the "great grey green greasy Limpopo River," where the elephant, you may recall, got his trunk). The structural movements in the southern African crust that have led to the formation of the present-day delta simply raised what is now southern Botswana, preventing the outflow of water from the lake from reaching the Limpopo drainage. The block faulting activity that led to the formation of the Okavango Delta diverted the Chobe and Zambezi Rivers, forcing them to make dramatic right-angle turns, eventually merge, flow over Victoria Falls, and find the sea hundreds of miles farther north than the Limpopo. The lake grew salty and eventually dried up.

Evidence of that ancient lake is everywhere in northern Botswana, just below the delta itself. Old shorelines stick out. Vast expanses of salt flats—reminiscent of the Bonneville Salt Flats in Utah, themselves relicts of an Ice Age lake (Lake Bonneville) many times the size of the present Great Salt Lake—lie as mute testimony to this once great body of water. The most extensive of these are the Makgadikgadi Pans, which still hold water in the rainy season, attracting what is left of the once great herds of wildebeest and zebra of the Botswanan Kalahari. Humans—our ancestors and ourselves—have lived in the region for at least 300,000 years, leaving a sequence of old, middle, and new Stone Age tools in and around the pans.

Another dramatic sign of this ancient lake and drainage system comes, oddly enough, from the discovery of diamonds in Botswana. South Africa's diamond industry was established in the mid-1800s, with the first diamonds coming from the farm of a Mr. DeBeers at Kimberley. By 1914, the "Big Hole" had already played out, though South Africa until recently remained the world's largest producer of diamonds. It now ranks fifth, while Botswana has moved into first position.

Three diamonds were all that had turned up for all the initial prospecting in Botswana, three diamonds recovered from a tributary of the Limpopo, which forms the southeastern boundary between Botswana and South Africa. Efforts to find the Kimberlite pipe—the actual source of the diamonds—had all failed because all the early prospectors had, naturally enough, assumed that the source pipe must lie somewhere within the existing Limpopo drainage.

They were wrong, but it wasn't until the mid-1960s, when Gavin Lamont, a seasoned DeBeers Consolidated Mining Company field geologist, tackled the job and solved the problem of the missing diamonds. Lamont, the story goes, had been much impressed by the theories of South African geologist Alex DuToit, one of the earliest and foremost exponents of continental drift—the forerunner of our modern theory of plate tectonics—the notion that, over the eons, continents had changed their positions vis-à-vis one another. In the early decades of the twentieth century, continental drift was taken seriously by only a handful of prominent geologists (Du-Toit and the German Alfred Wegener chief among them) and was generally laughed off as a schoolboy's fantasy inspired by the evident fit of the coastlines of western Africa and eastern South America. How little the experts knew!

Not only did Lamont sympathize with DuToit's vision of a restless Earth, but he also took quite seriously DuToit's conclusion that the southern African crust had been "warped" along a line extending from Zimbabwe (Rhodesia to DuToit) down through southern Botswana some 20 million years ago. If DuToit was right, thought Gavin Lamont, the basic configuration of the drainage of all of Botswana must have changed drastically. In a move that epitomizes the very best scientific procedure, Lamont then predicted that he would discover the source of the diamonds much farther north than anyone had so far looked. It took several years of painstaking, arduous field exploration, compiling bag after bag of sand, each of which had to be sifted and examined for telltale minerals that would give away the presence of the mother lode, the long-sought Kimberlite pipe. Patience was rewarded in 1967 at a site named Orapa, on the southern edge of the Makgadikgadi Pans far to the north of the Limpopo.

Seldom has theory, coupled with keen field work and observation, yielded such fantastic economic rewards. Lamont discovered a second Botswanan diamond source in 1971—this time with the aid of termites, which had brought the telltale mineral grains up from several hundred feet below the surface (this is not the last we shall hear of the impact of termites on the Kalahari-Okavango systems). When the British Protectorate of

Bechuanaland gained its independence, becoming Botswana in 1966, its 330,000 inhabitants ranked among the poorest in the world. More than anything else, the discovery of diamonds has sharply raised the per capita income in Botswana. DeBeers splits its proceeds, retaining 25%, while the Botswanan government receives 75%. More graphic a confirmation of a scientific theory—and the former existence of that giant lake and its outlets to the south through the Limpopo system—would be hard to imagine.

DYNAMICS OF THE KALAHARI AND OKAVANGO ECOSYSTEMS The game of life is
played out in the local ecosystems in each and every corner of the globe, from the Arctic tundra to the tropical rain forests, from sandy beaches to the reaches of the oceanic deep. Everywhere energy flows through the system, as plants convert solar energy to sugars, to be eaten by an army of insects and other invertebrates, on up through the great herbivorous mammals who, 65 million years ago, inherited their plant-eating niche from the now-defunct dinosaurs. Then there are the carnivores, the animals that prey on the plant eaters. Africa is virtually synonymous with the "big hairies," including herbivorous buffalo and elephants, to be sure, but especially conjuring up images of the big carnivores: wild dog, cheetah, leopard, and especially lion. Smaller carnivores abound as well; birds, lizards, and prodigious numbers of snakes, spiders, scorpions, and insects also are busily engaged consuming other kinds of animals. Lastly, but most crucially, come the humble agents of decay: fungi, bacteria, and protozoans, essential to the critical task of recycling organic debris and adding nutrients to the mix so that the game of life can continue, endlessly, generation after generation.

The Okavango-Kalahari ecosystems are a quilt-work of intersecting habitats. Flying north from Gaborone, Botswana's capital near the South African border, at an altitude of 300 meters, the acacia scrublands and grasslands of the southern and central regions immediately hove into view. Here, in its easternmost expression, the Kalahari is technically not a true desert at all, receiving (depending on locality) anywhere from 150 to 500 millimeters of rain a year (true deserts receive less than 100 millimeters). Farther to the west, primarily in Namibia, the climate is drier, and the land conforms more closely to more conventional, stereotypical images of true deserts.

Botswana's Kalahari environment is itself a patchwork of habitats, with plenty of open grasslands supporting—until very recently—huge herds of wildebeest, hartebeest (a related species of antelope), and zebra, with lesser numbers of other antelope, giraffe, warthog, jackal, hyena, and various cats, most certainly including lion. Several species of thorny acacia (a species of which was called "fever tree" by early European settlers because they mark the southern limits of the African malarial regions) form stands in the grasslands, creating a scrubland ideal for both grazers and browsing species. Giraffes are particularly fond of acacias, finding the prodigious thorns no obstacle at all. Other species, such as the ubiquitous impala, can also browse these trees. Impalas, which are antelopes, are the quintessential ecological generalists—jacks-of-all-trades that browse and graze a wide variety of grasses, shrubs, and low trees, and that are consequently equally at home in open grasslands, scrublands, open woodlands, and dense stands of forest. They do require water more than some desert-adapted species (such as the springbok, the southern African equivalent of East Africa's Thomson's gazelle, and, as such, the only true gazelle in the region). With plenty to eat, but with water a limiting factor, impala are more numerous in the wetter regions to the north, in and around the Okavango Delta, than in the drier regions of the central Kalahari.

Approaching Maun in our Kalahari flyover, the salt pans of the Magkadikgadi loom off to the right. Directly below, dense stands of Mopane woodland appear. Mopane is an important tree in southern African ecosystems: Its leaves, which close up along a fold line, have an extremely high protein content, making them the favored food of elephants, baboons, and many other species. The wood is very hard and tough and is the slow-burning fuel of choice for the traditional African cooking fire, which consists of two or three logs burning on end, radiating away from the flames.

With plenty of moisture in the soil, the mopane leaves are green, and the forest is alive with birds and an abundance of larger game. Arnot's chat, a sprightly black and white bird ever on the lookout for insects, is a common denizen, as are woodpeckers, several species of hornbills, warblers, and dozens of other bird species. Outside the delta, though, the dry season turns the green mopane leaves a coppery color; the leaves collect on the ground below the trees, and from the air these dry mopane woodlands are a desolate sight. You can walk, even drive, for hours through these dry mopane woodlands without catching a glimpse of a single mammal, or hearing the call of a single bird. It is as if these mini-ecosystems shut down, hibernating through the dry season in wait of the next burst of rains to bring life teeming back once again to their midst.

Approaching Maun, the Boteti River hoves into view. The Boteti used to flow year-round, the single outlet of the Okavango Delta that reached its end, finally, at the Mopipi Pan, a small salt pan just southwest of the Makgadikgadi, now used as a reservoir to slake the water needs of the diamond operation at Orapa. The Boteti hasn't run for several years, since the onset of the present drought in the late 1980s, but there is always, nonetheless, some pooled water present here and there—vital for the needs of wildlife and, in recent years, for the cattle that have effectively replaced antelope along the Boteti's course.

Approaching Maun, still at 300 meters in the air, the rondavel huts and bomas (corrals) of local Botswanans appear, and in addition to the usual signs of civilization, there are the unmistakable barren dusty signs of overgrazing by domestic livestock. Dry as the region just outside the delta may be, it would still be covered with grasses, yellowish in the dry season, verdant in the wet. Instead, light gray soil, a fine-grained clayey silt, is all that meets the eye.

Once past Maun, the scene once again changes dramatically, for we have entered the complex of habitats and vegetation zones that together form the Okavango Delta environment. The grasslands are intact, and look greener—even in the dry season—than they did in the Kalahari. Water springs into view, as a welter of lagoons, streams, and side channels develop. Along their sides, trees appear, forming gallery forests that ring the margins of the main channels. There are 80 *common* tree species in the Okavango Delta. Among them are jackal berry trees, whose fruit appeals to elephants, baboons, and humans. The sausage tree—so named for its half-meter-long seedpods that dangle from thin stringy stalks—produces a chemical substance recently shown to be an effective agent against some forms of skin cancer. There are also two species of palm tree, including the "real fan palm," which bears a tennis ball-sized "ivory" fruit much beloved

[Figure 2] *A Pel's fishing owl taking a fish at night. Like the unrelated osprey (a fish hawk), these owls have spicules on their feet to help them hold on to their prey. Owls have soft feathers at their wing tips, enabling them to fly silently through the night air. Pel's fishing owls are sometimes seen by day, when flushed out by humans moving through the riverine forests of the Okavango Delta. I saw my first one from the back of an elephant.*

by elephants and humans alike. This species, together with the wild date palm, are often home to the African palm swift, the red-necked falcon, and several other species of breeding birds.

Islands abound in the southern reaches of the Delta. The largest ones, with Chief's Island by far the dominant, house the greatest variety of habitat. Typically, reed beds fringe the shores of the streams, but majestic gallery forest also comes to the water's edge on the larger islands. The forests house a rich assortment of birds, most notably Pel's fishing owl, a magnificent large brown bird that takes its prey by plunging feetfirst to the very bottom of the pools flanking the forest's edge. [Figure 2] The beautiful, small chobe bushbuck hides in these dense forests.

Farther inland, more open mopane woodlands (typically green the year around because of the continual presence of water) and acacia scrubland are occasionally present. Invariably, though, there are open grassland savannas, sometimes approaching quite close to the water's edge and generally extending onto and covering the interiors of the islands.

It is on these islands that the closest setting to the primordial African—meaning the primordial *human*—ecosystem persists. Here the ancient rhythms of life are still being played to their fullest. The sheer variety of mammal species in itself tells the tale. Near the water's edge, large herds of red lechwe, a medium-sized, tawny antelope adapted to grazing riverine grasses and escaping wild dogs and lions by dashing through the lagoons and reed beds on toes well-adapted for marsh muck, are the most common sight. [Figure 3] Reed buck, a smaller, tanner species, form smaller groups that haunt the taller grasses, again not too far from the

[Figure 3] *Deep in the Delta. The sitatunga, with its splayed toes, is a common but rarely seen antelope that inhabits dense stands of papyrus. The sitatunga shuns the more open waters, where African jacanas trot on top of water-lily leaves.*

open water. Larger waterbuck, a gray species with prominent white rings around their derrières—a species whose flesh is sometimes said to be repugnant to both man and beast—also, as their name implies, like the water, though they can also be seen ranging farther inland than red lechwes and reed bucks.

Impala are everywhere. Their distant cousins, the tsessebes, however, are more specialized, preferring the open grasslands and tree-dotted savanna grasslands of the island interiors (there are plenty of water holes that persist virtually year-round on these islands). Tsessebes (the same species, *Damaliscus lunatus*, as eastern Africa's topi) are ungainly beasts, and it usually surprises safari-goers to hear that they are the fastest antelope species in Africa. A grayish-brown, tsessebes cluster around the grasslands in generally smallish groups of 5 to 10 individuals.

Blue wildebeest (the same species, *Connochaetes taurinus*, as the East African gnu that everywhere forms such huge, migratory herds) are close relatives of the tsessebes. They abound on the grasslands of the Okavango, almost always within sight of small herds of Burchell's zebra. Specializing on different grasses, the intimate association of wildebeest and zebra reflects the mutual benefit each derives from their differing abilities to detect predators.

The list of antelope species is long: little steenboks and slightly larger gray duiker fall on the smaller end of the size spectrum, while the magnificent kudu (the horns of the male are used as natural trumpets

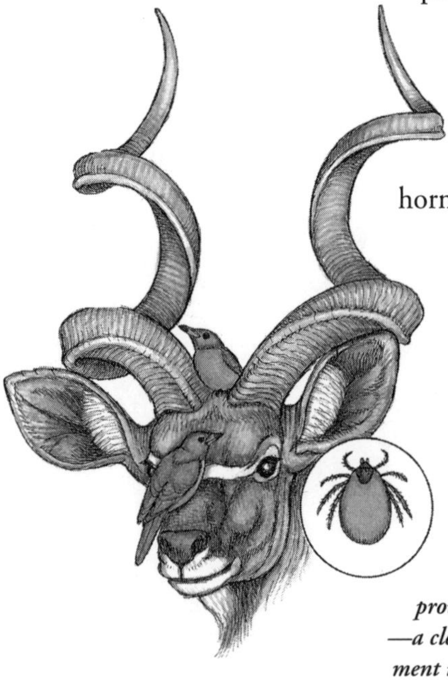

[Figure 4] *Tick-eating oxpeckers on board a magnificent male greater kudu. Oxpeckers are related to starlings. The two species of oxpeckers provide a valuable service to their mammalian hosts —a clear example of "commensalism," where the arrangement is beneficial to antelope and oxpecker alike.*

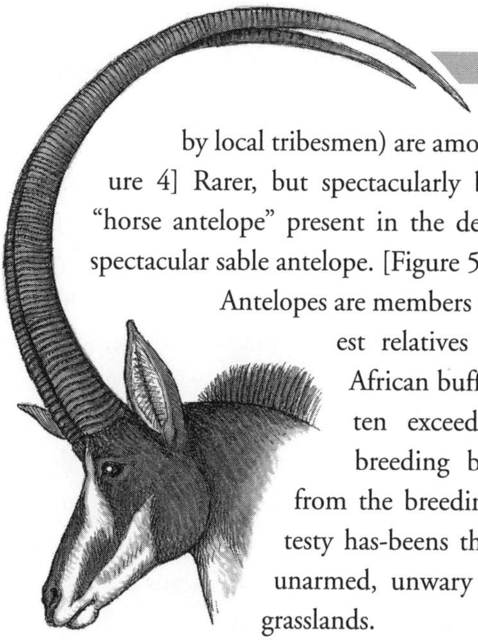

by local tribesmen) are among the larger of all antelopes. [Figure 4] Rarer, but spectacularly beautiful, are the two species of "horse antelope" present in the delta, the roan and the even more spectacular sable antelope. [Figure 5]

Antelopes are members of the cattle family, and their closest relatives on the Okavango savannas are African buffalo, which occur in herds that often exceed 1,000 individuals. The older breeding bulls are systematically excluded from the breeding herds, forming small knots of testy has-beens that are the most serious threat to unarmed, unwary foot travelers on the Okavango grasslands.

Giraffes, favored food of lions in the Okavango, are abundant and grace the woodland margins and open plains in small herds. Both species of rhino (black and white) are scarce— virtually gone to the poachers, who still take their horns to feed the insatiable Asian aphrodisiac market. In sharp contrast to the plight of the rhinos, though, elephants are undergoing a population explosion in the delta, as we will discover in greater detail below.

[Figure 5] *A sable antelope. Together with its close relative, the roan antelope, sable antelope are among the most beautiful and imposing of the Delta's mammals. A third species, the bluebuck, also belonged to this "horse antelope" group—but was driven to extinction in the 1790s.*

Hippos thrive in the delta's waterways. Lurking in the deep channels, sometimes rearing up suddenly to challenge a fast-moving motor boat laden with tourists, hippos tend to accumulate in larger herds in the more open waters of the side channels, adjacent to the interior grasslands of the islands. There they share space with the ubiquitous Nile crocodile—more or less at peace with one another, though crocs will take young hippos, and adult hippos will kill even large crocs when threatened.

Warthogs are all over the place; their cousins, the bushpigs, are harder to find, preferring dense cover and, in any case, being primarily nocturnal creatures. Three large rodents finally round out the long list of mammalian herbivores: the African porcupine, the tree squirrel, and a large hopping rodent called the springhare, rarely seen by day, but out in droves at night.

The list of carnivores is almost as long as that of the herbivores. Six species of cats run the gamut from the house-cat-sized African wildcat (probable ancestor of modern domestic cats) up through lions. Servals are slightly larger, spotted cats; caracals are the equivalent of the North American lynx. Cheetahs come next in size; predators of the open plains that chase down their prey in wind sprints, cheetahs nonetheless hold their own in the smaller expanses of grasslands of the delta. Leopards, true ecological generalists like impalas, are at home in deserts, mountain ranges, and tropical rain forests. No surprise, then, that leopards are also abundant in the Okavango. At the top of the heap, of course, are the lions that likewise thrive in goodly numbers on the region's savannas and scrublands.

Hyenas and civets are close relatives of cats. Spotted hyenas are common, often bringing their maniacal laugh into camp at night. Much harder to find are the smaller brown hyenas, as are aardwolfs, hyena relatives that eat only termites. The African civet and small spotted genet round out the roster of catlike mammalian carnivores. Not to forget the three species of mongoose, weasellike carnivores that nonetheless are more closely related to the civet branch of the mammalian carnivores.

But there's still more: Four species from the dog family roam the savannas and marginal woodlands. Smallest is the bat-eared fox, followed by two species of jackal (black-backed and side-striped). Surprisingly common is one of Africa's most endangered species, the wild dog [Figure 6]. Recent research in the Okavango has shown a tendency for the wild dogs there to hunt singly, rather than in packs that dominate wildlife films from East Africa.

There is also a species of otter in the waterways, and, back in the grasslands, the honey badger, a rough customer known to have attacked even elephants, with a reputation of going for the groin when provoked. The skunklike striped polecat finally exhausts the list of mammalian carnivores.

African anteaters include the antbear and the scaly pangolin. Both

[Figure 6] *A wild dog—member of the species* Lycaon pictus. *I once encountered a pathetic pile of desiccated skin, hair, and bone in a dusty plain in the Delta. The naturalist guide said "wild dog kill"; I asked "when?"—expecting him to say "weeks ago." Instead, he replied "Last night."*

species stick out like sore thumbs in an evolutionary sense: Neither has any other close living relatives. They are "living fossils," whose closest kin lie buried deep in the fossil record of earliest mammalian history.

Which, with the conspicuous exception of the many species of mice and bats, leaves only the primates, our closest kin. Aside from ourselves, three primate species live in the Okavango ecosystems. Vervet monkeys occur in small bands, eating an omnivorous diet of fruits and insects. More conspicuous are the troops of chacma baboons. Omnivorous, like vervets, baboons will take small antelope to supplement their diet of scorpions, insects, and, of course, fruit. Least conspicuous are the nocturnal lesser bushbabies, a species of lower primates that can jump over 10 meters seemingly effortlessly.

That's just a lightning-quick tour of the commonly encountered mammals alone. All told, there are 164 mammalian species known from the delta. Add to them, 38 species of amphibians (frogs and salamanders), 157 species of reptiles (snakes and lizards, including the meter-long monitor living along the waterways), and the great wealth of bird life (540 species), and you have a very rich set of players in the Okavango ecosystems indeed.

And that's just for starters. The insects are incredibly diverse (at least 5,000 species!), as are the spiders, scorpions, and sunspiders. Butterflies, walking sticks, grasshoppers, and beetles abound. Ants are very diverse. Then there are the termites, in many ways the backbone of the Okavango's terrestrial ecosystems, whose story is so complex, it is told as a microcosm of ecosystem dynamics below. Add to this melange of animal life, the 100 or so species of trees, many additional species of grasses—all told, at least 3,000 species of plants—plus the microbes in animal intestines, in plant tissues, and free-living in the soil, and the total riot of life in the Okavango emerges, adding up to churning ecosystems whose vibrancy is somehow greater than the sum of their parts and that cannot be done justice by simply listing all the species found there.

If the savannas, woodlands, and scrublands of the delta's islands are the consummate vestiges of primordial Africa, there is still more to the complete picture of the Okavango habitats. Forsaking a plane for a motorboat, a cruise along the main channels northward into the heart of the delta yields a very different picture from that seen in a flyover. For one thing, in the eastern reaches at least, the main watercourses are lined with tall reeds, peppered with small stands of papyrus. Even standing in the boat, you cannot see beyond the thick wall of grasses and papyrus, except in places where the watercourse widens, or where there has been a recent fire. As you go northward, the papyrus increases in frequency, until it is the dominant and, ultimately, the only plant lining the waterways. This plant, with its long, tough green stem and frizzy top knot, has given us the very word "paper."

The Gcodikwe heronry, on one reed-and-papyrus-choked island in the very heart of the delta, is perhaps the largest of all the many breeding sites for the Okavango's herons, egrets, ibises, storks, cormorants, and darters. Here, fiercely ugly marabou storks share space with much more comely yellow-billed storks. The saddle-billed stork, largest of the region, is one of the most beautiful birds in the world.

Traveling farther north, the islands eventually all but disappear, and the entire delta is one vast papyrus wetland. Here is where the most elusive of all antelopes is most common: the sitatunga, an animal whose hooves are splayed even more than the red lechwe's, enabling it to live virtually its entire existence in the reed beds and papyrus swamps.

Turning the boat around and returning southward along the western side of the delta, there is one new twist to the landscape: Gone, for the most part, are the dense stands of tall papyrus and reeds that line the channels to the east. The grasslands come right down to the banks, allowing an unimpeded view of the wildlife and an opportunity for larger crocodiles to sun themselves at the river's edge.

Despite the illusions of casual Western visitors that they are in some kind of a game park, and despite all the damage already done by human encroachment, the Okavango's ecosystems are astonishingly intact, incredibly complex, yet at the same time evincing a kind of simplicity that only systems as old, as well broken in, can possibly show. The Okavango is the real thing.

TERMITES IN THE OKAVANGO SCHEME OF THINGS It is difficult to capture the dynamism of local

ecosystems simply by listing the species, the players in the ecological game of life. The intricate web of energy flow and matter exchange starts with plants trapping sunlight, storing the energy in sugars that they—and those that eat them—then utilize to grow, maintain their bodies, and reproduce. Focusing on one intermediate step in this energy web—the role that a single species plays in the Okavango grasslands and scrublands—helps capture the complex interdependence of all living components of the ecosystem.

Consider the role that termites play, especially in the arid Tropics where they are so abundant. Without termites in these ecosystems, there simply would be no decay of cellulose, the predominant component of plant tissue. Without decay, the system would rather quickly grind to a halt, as nutrients would remain forever trapped in dead wood and stems, and a thick carpet of dead plants would quickly begin to clog the landscape. Termites, as all homeowners know, can eat and digest cellulose, though how they do so is in itself an interesting ecological story. Termites themselves cannot digest cellulose directly. Just as we depend on a rich flora of microbes in our stomachs and intestines for complete digestion and absorption of nutrients, termites house, in their hindgut, a truly amazing array of microbes, including various bacteria and protozoans. These microbes actually convert the cellulose to utilizable energy—an example of ecological mutualism beneficial to host and microbe alike, but with vast implications for the ecosystem at large.

As if the digestion of cellulose itself were not a sufficiently monumental service paid to the ecosystem, it has recently been discovered that the microbial ecosystem of the termite hindgut also "fixes" nitrogen, extracting and storing it in a form usable by other organisms. All animals require nitrogen to manufacture proteins and DNA and have no means to derive the substance from the inorganic world, even though nitrogen forms 79% of the air that we breathe. For years it was thought that only the bacteria living in association with the roots of leguminous plants (such as peas) performed this absolutely critical task of nitrogen fixation, but now it seems that nitrogen also enters the organic cycle through the efforts of the microbial microecosystem developed in the hindgut of termites.

Macrotermes michaelseni is the most conspicuous termite in the Okavango region—not that the insects are themselves usually seen except when swarming to fly off to found other colonies. Rather, it is their

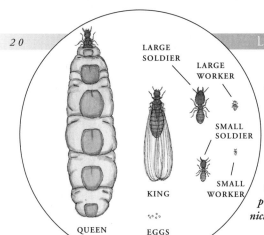

LARGE SOLDIER

LARGE WORKER

SMALL SOLDIER

SMALL WORKER

KING

QUEEN

EGGS

SOCIETY OF TERMITES

Macrotermine termites are divided into four castes—large soldier, small soldier, large worker, small worker—plus the queen and king. The primary queen lives for an average of 10 years, laying about 30,000 eggs per day, or 10 million per year. The ratio of caste members and the social activity of the colony is regulated by a wide range of reproductive, chemical, nutritive, and communication strategies.

FUNGUS GARDENS

Excrement of termites is used to build the delicate shelves on which the fungus combs grow. The fungus degrades the excrement which contains partially decomposed grass cellulose. The termites constantly eat mature portions of the fungus which is further degraded by microbes in the termite's hindgut. In this way the cellulose is almost completely metabolized by the termites.

THE ROYAL CELL

The queen and king are housed within a large protective mass of clay generally located on the floor of the mound cavity. The cell contains numerous small holes through which soldiers and workers move freely to feed the royal pair and remove eggs to the mound nursery.

WORKER CULTIVATES FUNGUS

WHITEFRONTED BEE EATER

AIR FLOW

FUNGUS COMBS DEVELOP IN THE MOUND

CELLAR

[Figure 7] *The ecologically important termite* Macrotermes michaelseni. *The large mounds of colonies of this termite species dot the landscape in both the Kalahari and the Okavango Delta. A mature mound can contain over 5 million living individuals at any one time. The termites live their entire existence in the closed environment of the mound and the tunnels that radiate from it to the main source of their food, the Savannah grass,* Urochloa trichopus. *Macrotermine termites are unique in their cultivation of fungus within the mound. The basidimycete fungus* Termitomyces *is known to grow only in the mounds of these termites, and each species of fungus is specific to a specific species of termite.*

EUROPEAN SWALLOWS
SPEND SOME OF THE
YEAR HUNTING TER-
MITES

PREDATORS
While aarvarks, aardwolves, mongooses, European swallows, and whitefronted bee eaters all eat termites, it is the driver ant, Megaponera foetens, *which is the dominant predator of macrotermine termites.*

LARGE
DOUBLE
MOUND

AARDVARK

AIR CONDITIONING SYSTEM
The elaborate system of vents within the mound's central cavity serves to regulate the temperature and humidity such that an internal microclimate is maintained between 25° and 30° C and at a humidity of 92% (±4%) throughout the year, independent of external climate. Neither the termites nor fungus can live for long outside of this range.

enormous mounds dotting the landscape virtually everywhere that are so conspicuous. [Figure 7] The mounds are often over 3 meters high, and function, so far as the termites are concerned, as air conditioning structures and essential protection from their many enemies.

These termites are loaded with fats and proteins, and many species find them delicious. African peoples have consumed termites for millennia. School kids, finding a fresh termite swarm, will fall on them and gorge themselves before going off to tell their friends of their discovery (or so a Ugandan college roommate of mine once told me—at the same time professing disgust at the mere thought of anyone eating shrimp, crabs, or lobsters; *chacun a son goût!*).

Termites are on the menus of many other species as well. Their most persistent and formidable enemies are driver ants, which is why termites labor so intensively to keep their gray-clay mounds impervious to even the smallest of invaders. But they also are beloved by many larger animal species, especially aardwolves and aardvarks, as well as many omnivorous species, including certainly chacma baboons and vervet monkeys, that happen to chance on them.

So much is fairly typical—if dramatically so—of the role of a single species in an ecosystem: *Macrotermes michaelseni* is important because of what it eats, how it cycles nutrients through the system, and because of what eats it. (Unlike other termites, cellulose digestion in *Macrotermes michaelseni* colonies is actually handled by cultivation of fungal farms inside the mound itself rather than by microbes living in their hindguts). But position in the food chain–energy web hardly exhausts the importance of these termites in the Okavango scheme of things. For these termites are literally master builders, and their mounds far transcend their original purpose of providing shelter for their builders.

To begin with, unused termite mounds provide homes to other species; aardvarks, hyenas, and many others seek out tunnels in termite mounds. Holes in sides of termite mounds cut through as they lay exposed on river banks are nesting sites for swallows and whitefronted bee eaters, and the deadly black mamba is a frequent denizen, as it seeks out small rodents hiding in the abandoned channels.

But termite mounds play even larger roles than providing shelter for other species. What little topographic relief exists in the Okavango is crucial to the maintenance of its complexity. The main river courses have, of course, the deepest beds. They are flanked by shallow marshlands and wetlands. When the floods come, though, the waters overstep their banks and flood low-lying grasslands, greatly reducing, for a time, the space open to the dryland browsers and grazers. Many tree species, too, cannot tolerate periodic inundation. Thus, without some higher ground to provide year-round refuge from the floodwaters, the Okavango ecosystem complex would not be as rich as it is. Once again, we look to termites—especially *Macrotermes michaelseni*—to understand why there is any high and dry ground whatsoever in the delta.

Termite mounds frequently have trees growing out of them—trees dependent on both the slightly higher ground that a termite mound automatically creates, and perhaps as well the moisture that termites bring up from the depths. Bring in a tree, and the accumulation of leaves, seeds, dust, and animal droppings starts an accumulation of soil. Termite mounds are the very cores of small islands sticking out among the floodwaters, islands that will grow with the addition of more trees and shrubs. The islands grow by accretion, and termites play a critical role in the very creation of permanently dry grassland, scrubland, and woodland habitat in the Okavango.

Warthogs play an equivalent, if opposite role: The shallow excavations warthogs make while rolling in a mud puddle is often expanded when elephants seize the opportunity to mud bathe there, too. Enough elephant wallowing and pretty soon there is a large depression, hardened and clay-lined as it bakes in the sun during the dry season, forming a perfect cachement basin when the waters return. Thus, small wallows can quickly grow to sizable ponds—perhaps eventually with permanent water—homes for crocodiles waiting for the antelope to come to drink. Though it is easy to see how animal species are adapted to their environments, it is less obvious how these species act to create that environment in the first place.

The interplay between species in systems like the Okavango far transcends who eats whom. That the daily routine of humble little termites eventually determines where elephants, buffalo and lions will roam gives us some handle on what the complex interplay between species in tropical ecosystems is really like. No matter how we look at the Okavango and its surrounding Kalahari systems, we see that, up until now at least, these systems are healthy, vibrant, alive. So much is clear—but how do we really know the delta resembles Eden, the environment from which we originally came?

PART II: THE OKAVANGO AS PRIMORDIAL EDEN *When,*

in earliest recorded history, the upper and lower reaches of the Nile Valley were united to form the first Egyptian dynasty, the lotus (water lily flower) came to symbolize the Nile Delta, while the papyrus became emblematic of the inland, up-river region. This double icon persisted for thousands of years as the dual floral symbols of the Egyptian state. Unification came 5,000 years ago. Two thousand years later, stands of papyrus were still common in Egypt, becoming biblically famous as the bulrushes in which the infant Moses was hidden.

The famous Egyptian tomb paintings have revealed much about life in the Middle East thousands of years ago. Although the pharaonic tomb paintings tend to dwell on religious themes, tombs of the nobles—the literate class of scribes, tax collectors, and army officers—often depict scenes from daily life. At the ancient burial site of Saqqara, just a few hundred meters from King Djoser's Step Pyramid (the oldest known standing building in the world), single-floor rectangular burial chambers (mastabas) for some of the noble families of the earliest dynasties are filled with colorful paintings. There the game of the ancient Egyptian countryside runs as thick as it still does today in the Okavango Delta. Scenes of fishermen hauling nets laden with many species of Nile fish also show the hippos and crocodiles that were then so common among the papyrus and reed beds. Today, the only papyrus you can see in Egypt are a few plants cultivated just outside the Egyptian Museum in Cairo and a few isolated stands along the banks of Lake Nasser, at such tourist sites as Abu Simbel. Hippos are long since gone, as are the fearsome Nile crocodiles; Sobek, the Crocodile God, was worshipped—especially in swampy places! Thoth, the God of Wisdom, was represented by the sacred ibis, tens of thousands of whose mummified bodies have been excavated in tombs in central Egypt. The sacred ibis has now entirely disappeared in Egypt, though it is still a common breeding species at the Gcodikwe heronry in the midst of the Okavango Delta.

The Okavango environment was clearly much more widespread just a few thousand years ago than it is today. What happened in Egypt was a one-two punch of climate change and human impact: Northern Africa has been drying up for the past 10,000 years, and the Sahara has grown at the expense of formerly better-watered plains. Agriculture and the growth of political states has also spurred rapid human population growth. Even in the early days of the Old Kingdom, human population was on the rise, and the spiral of conversion of lands for agricultural use—raising more food, feeding more people, supporting higher birthrates, leading to the demand for still more food—was well under way. Canals were built and waters diverted, a process culminating in the construction of the two Aswan dams of the twentieth century. Add to that overfishing and the extermination by hunting of hippos, crocodiles, antelope, and birds, and you suddenly understand why no vestiges of the Nile Valley's wild, natural ecosystem remain.

Except in the Okavango. To look at the Okavango is to look at the primordial Egyptian setting, at the way things were at the dawn of agriculturally fueled settled existence, civilization. To look at the Okavango is to see still more, to stare still more deeply into our collective human past. To look at the Okavango is to see our very own birthplace. The Okavango is the environment from which we all collectively sprang, and the place where our ancestors—species not yet totally human—were born, played out their lives, and were ultimately replaced by none other than ourselves.

The Okavango Delta, recall, lies in the southwestern most, and most recently formed, segment of the great East African Rift Valley system. Two thousand miles to the northeast, Tanzania's Olduvai Gorge is one of the most famous sites for the discovery of some of proto-humanity's earliest fossilized remains, discoveries made famous by the intrepid and colorful Leakey family. Along with these precious relic bones have come the earliest stone tools, evidence of humanity's earliest flirtation with material culture, so vital to the ultimate ecological success of our species, *Homo sapiens.* Over the past 30 years, the initial successes at Olduvai have prompted an explosion of fieldwork. Important remains have now been found all the way along the Rift Valley, from Ethiopia far to the north (where Don Johanson's team recovered the famous "Lucy"), down through the wilds of northern Kenya (where Richard Leakey made important finds in the '70s and '80s along the shores of Lake Turkana). At Laetoli in Tanzania, Ron Clarke excavated Mary Leakey's exciting find of 3.5-million-year-old footprints, proof that our ancestors were upright and walking on their hind legs long before the hominid brain was anywhere nearly as fully developed as in modern human beings.

Africa is the place where the earliest episodes of the human evolutionary drama unfolded—after, that is, our lineage split away from the line that led to modern-day chimpanzees and gorillas some 6 million years ago. We associate human evolution with the rift system because that's where the fossils have been predominantly found. The lakes and streams that filled the valley to the east and north of Botswana 2.5 million years ago accumulated thick deposits of sands, muds, and volcanic ash—layer upon layer of what are now thick sequences of sedimentary rock. The bones—of humans and the entire range of mammals, reptiles, birds, and fish, not to mention the shells of freshwater invertebrates, and, crucially, the pollen grains of then-resident plants—were routinely trapped in this pileup of mud, the prerequisite for forming a rich and dense fossil record. The Rift Valley not only was part and parcel of our ancestral home, it was ideally suited to accumulating a record of our 6-million-year history there.

Other settings, places just outside the Rift Valley system proper, have also preserved a record of this ancient human evolutionary activity. High in South Africa's Transvaal—a region of rock billions of years old, harboring that country's famous gold and diamond deposits—lie a series of limestone caverns that have also produced a rich fossil record of proto-humans. Debris falling into caves, including the bones of ancestral human species apparently killed by leopards (or, some now say, black eagles), accumulated in rubble piles on the cavern floors. The layering is not as neat and tidy as in the lake deposits of the Rift Valley, but diligent work beginning in 1924 has nonetheless revealed a story of early African human evolution that agrees very well with the story pieced together from the East African Rift Valley bones and stone tools.

Here, in a nutshell, is that story. Five million years ago, the Mediterranean dried up in the amazingly quick span of only a few thousand years. Gibraltar had closed through a spasm of Earth movement that brought the African and European plates together. Even today, the Mediterranean absolutely depends on the constant influx of Atlantic waters. Because the evaporation rate is so high in the eastern reaches of the Mediterranean, and because not enough freshwater falls as rain or comes in from the Rhone, Danube, and Nile (the three major rivers dumping into the Mediterranean), a dense salty brine forms in the east, sinking and flowing westward, escaping from the Mediterranean into the Atlantic at Gibraltar. When Gibraltar abruptly closed 5 million years ago, no "fresh" marine waters could enter, nor could the brine escape, and the sea quickly became a very deep, and, for a time, very dry, salt pan.

The quick evaporation of the Mediterranean was spurred by a major change in global climate that occurred about the same time, a spasm of global cooling only detected in recent years. An early harbinger of the cooling event that was to start some 2.5 million years later, the effect of this early spurt of global cooling on African ecosystems was stark: With the lower temperature came drier conditions, and the wet forests began to give way to more open savannas.

In this mix of forest and an emerging savanna habitat, initial critical stages of human evolution took place. Foot bones from the lake deposits of Ethiopia's Rift Valley show that the earliest human species to walk up-

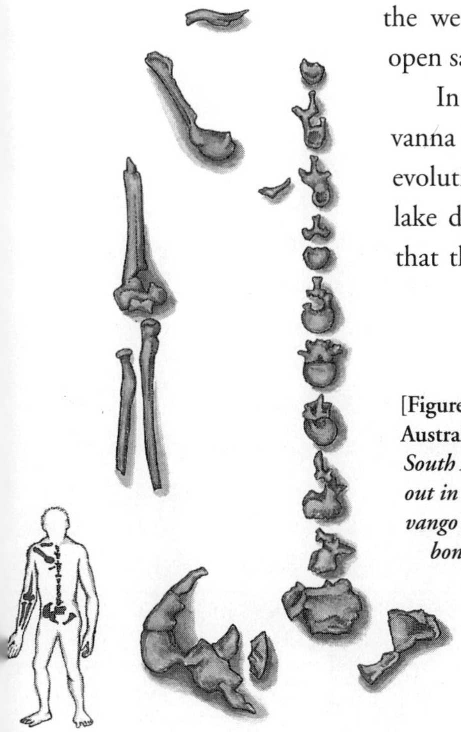

[Figure 8] *Partial skeletal remains of an individual* **Australopithecus africanus** *from Sterkfontein Cave, South Africa. Further exploration of Drotsky's caves, out in the Kalahari west and a bit south of the Okavango Delta, may also prove to have the fossilized bones of this and other ancient hominid species. I once saw a large leopard sticking his head out of the mouth of one of these caves—a starkly dramatic reminder of the way the bones of* **Australopithecus** *likely found their way into the caves of southern Africa.*

right (*Australopithecus afarensis*—Johanson's "Lucy") still retained vestiges of the ability to cling to branches.

It was the grasslands that tempted *Australopithecus afarensis* to come out of the trees, to traverse the newly developed plains in search of edibles: small mammals, perhaps carcasses, as well as nuts, tubers, and other plants. Walking upright presented several advantages. For one, vertical posture reduced exposure to the hot sun. Coupled with a greatly increased network of blood vessels on the surface of the brain, our ancestors developed an efficient means of staying relatively cool under the blazing African sun to which they were newly exposed on the grasslands. Standing upright as a 1.2-meter-tall hominid would also ease the task of keeping watch for predators in the unsafe tall grass.

Most critically, walking on your hind legs frees up your hands. Anyone who has watched baboons (as monkeys, split off from our own lineage at least 20 million years ago) has seen them walking along on all fours, stop-

[**Figure 9**] *African ecosystems—before and after the global cooling event some 2.5 million years ago. The first scene depicts a wetter and more densely forested climatic regime— while the second, a "snapshot" of African life some 1.8 million years ago, shows the spread of drier, less wooded grasslands at the expense of the older forests. Though the mammalian denizens of these African landscapes look superficially similar, extinction and evolution of species, plus migration in and out of eastern and southern Africa to keep pace with shifting habitats, means that most species present in earlier times depicted to the left are no longer present in the later scene.*

2.8 MILLION YEARS AGO

GIRAFFA

AUSTRALOPITHECUS   STRUTHIO

HIPPOPOTAMU

DEINOTHERIUM

HOMOTHERIUM

ping to pick up a berry with one hand while leaning on the other. To use both hands, a baboon must sit back on its haunches, to become a "man who sits on his heels" as the San people say. Not so with a bipedal human. The freeing of our hands, with our fully opposable thumbs, coupled with the stereoscopic (3-D) color vision we brought with us from our earlier days in the trees, made possible a fine level of manipulation of objects not given to any other creature, including our collateral kin in the primate world.

Thus, we think we basically know why human intelligence developed, and why our lineage set off on a course of increasingly culturally contrived approaches to making a living: surviving on the margins of woodlands and savannas; finding and securing food and eventually actively hunting; warding off predators; finding—and later making—shelter; using fire for warmth, protection and, ultimately, the preparation of foods. Hands were freed and imaginations spurred—all because *Australopithecus afarensis* evolved a bipedal gait in response to a climatic event, a burst of global cooling that transformed the African landscape 5 million years ago.

How climate induces bursts of evolution—and, critically, extinction— is seen much more clearly in the events of 2.5 million years ago, which are displayed in the fossil record of both the East African Rift Valley and the cave deposits of the Transvaal, just to the south of the rift system. Once again, beginning about 2.8 million years ago, a cold snap gripped the

2.5 MILLION YEARS AGO

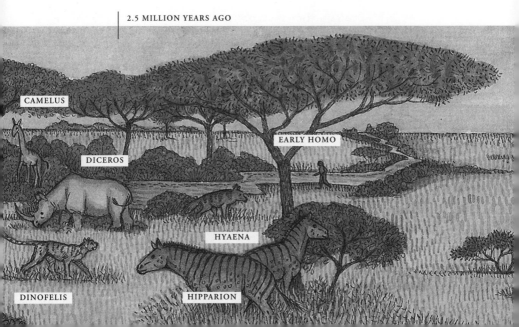

CAMELUS

EARLY HOMO

DICEROS

HYAENA

DINOFELIS   HIPPARION

world. Once again, the ecosystems of Africa were transformed as savannas spread even farther.

Australopithecus africanus—a 1.2-meter-tall creature with a brain little larger than that of a modern chimp, but nonetheless equipped with a human pelvis and near-human foot structure that are unmistakable signs of belonging to our ancestral lineage—lived in southern Africa from about 3 to around 2.5 million years ago. [Figure 8, see p. 27] Its environment seems to have been almost identical to the region around the larger islands of the Okavango Delta today. Recent analysis at Sterkfontein, the premier site for early human fossils in South Africa, reveals the long vines of lianas that used to hang down into the gloomy recesses of caves. The caves were surrounded by gallery forest, yet the presence of the species of antelope that are also found in the cave deposits shows that grasslands and running water could not have been very far away.

That these proto-humans were very much a part of the local African ecosystems of the time is starkly portrayed by the events that came swiftly on the heels of this cold snap 2.5 million years ago. After absorbing a steady decline in temperatures for one or two hundred thousand years, the African landscape suddenly changed, with a second spurt of growth of open savannas at the expense of much of the remaining woodlands. [Figure 9, see pp. 28–31] The vast plains of the present-day East African Rift Valley abruptly developed—plains very different in

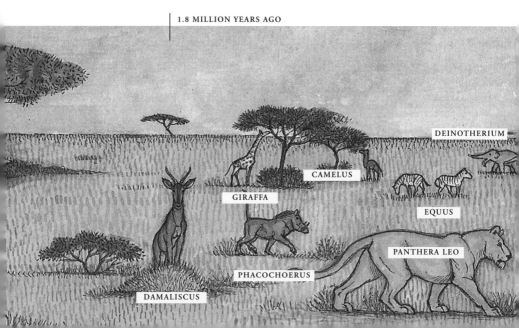

1.8 MILLION YEARS AGO

DEINOTHERIUM

CAMELUS

GIRAFFA

EQUUS

PANTHERA LEO

PHACOCHOERUS

DAMALISCUS

their sheer extent from the much smaller patches of savanna of the Oka-vango Delta region.

Two very different sorts of things happen when landscapes are disrupted as suddenly and radically as they were when the atmosphere began to cool just over 2.5 million years ago. The most obvious is that habitats simply move around. When, for example, the glacial pulses began a little over 1.5 million years ago, the tundra, northern forests, and plains of Europe and North America all moved farther south, keeping pace with the southward growth of massive sheets of glacial ice. When the ice melted back, as it did periodically, the tundra, forests, etc. simply moved north. As the vegetation moved, so did the animals. The net result is that climate change moves habitats around; Africa got its first antelopes during that *first* cold snap 5 million years ago, when grassland savannas began to take shape, the Mediterranean dried up, and Eurasian antelope species simply migrated for the first time into the African scene.

But what happens if a species cannot keep pace, cannot go with the flow, cannot continue to recognize familiar habitat elsewhere, or simply cannot move quickly enough to find the familiar habitats to which it is adapted, in which it must live? The answer is as simple as it is stark: These are times when many species simply cannot go on. They drop out. They become extinct. The 2.5-million-year-old fossils of the East African Rift Valley system are especially clear on this point: A large number of species

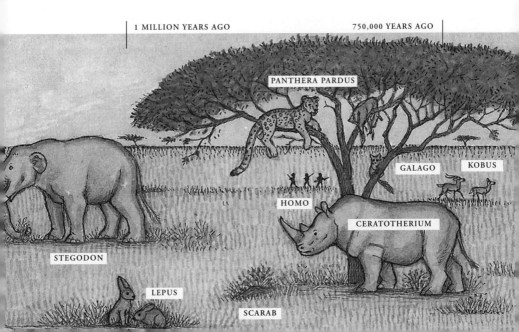

1 MILLION YEARS AGO   750,000 YEARS AGO

PANTHERA PARDUS

GALAGO   KOBUS

HOMO

CERATOTHERIUM

STEGODON

LEPUS

SCARAB

from all major groups (including pigs, hippos, antelopes, birds, predators) simply ceased to exist—unable to survive, either by moving to familiar climes elsewhere, or by changing to meet the challenges of the new environment.

Right around 2.5 million years ago, there was a major die-off of the African biota, a die-off that included our own ancestral species, *Australopithecus africanus*. But extinction, the attrition of species unable to track suitable habitat as it shifts locale in times of environmental change, is not the only outcome, the only long-term response to such periods of stress. The habitat disruption that comes with severe episodes of climatic change also fragments habitats, isolating some species into relatively small groups—the very conditions needed for rapid evolutionary response. Out of disaster comes—at least for some—bursts of evolutionary creativity. Just after 2.5 million years ago, a number of new antelope and other species suddenly appeared on the scene, products of rapid evolutionary diversification that was one element of the response to environmental change.

That's exactly what happened to our own ancestral lineage. We lost our ancient ancestor, *Australopithecus africanus*, but, on the other side of that 2.5-million-year-old environmental Rubicon, we find not one, but two separate lineages of ancient hominid—one that eventually led to us, the other of which was a specialized evolutionary sideline that was destined to flourish for a million years before succumbing to extinction. Because our lineage tells the very same tale of extinction and evolution surrounding that climatic divide 2.5 million years ago—a climatic event that had such a pronounced effect on all the other components of the African ecosystems of that time—we can be certain that our ancestors were very much a part of the local ecosystem.

One of these new species led directly to ourselves. Tellingly, the Leakeys long ago had spotted very primitive stone tools at a layer dating back to 2.5 million years. First by circumstantial evidence, now by actual fossils, we can say with certainty that these first tools were made by *Homo habilis*—a species with an enlarged brain—the first sentient, if not fully human, tool-making, and thus culture-bearing hominid species. More recent finds of fossils and tools from Sterkfontein in the Transvaal show that *Homo habilis* lived over a wide African range.

We don't know for certain what became of *Homo habilis*. We do know that its place was taken, just over 1.5 million years ago, by a different species, *Homo erectus*. Like its predecessor, *erectus* was the evolutionary product borne of a cold snap—this one the first of the four great glacial pulses in the northern hemisphere that had such a chilling effect on the entire globe.

With an even bigger brain than *habilis* (though not yet as large as our own), and standing about as tall as modern humans, *erectus* continued the trend toward utilizing culture as a means of making a living. *Homo erectus* knew fire, and, when the second great pulse of glaciation hit, just under a million years ago, *erectus* actually left Africa, traveling north and invading the frozen expanses of Eurasia, very likely to take advantage of the huge herds of Ice Age mammals that had just recently evolved.

That was merely the first of several waves of out-of-Africa hominid treks, expansion into new ecosystems so very unlike the original Okavango-like Eden, where our early ancestors were spawned. Their growing capacity for culture enabled our ancestors to expand the range of habitable climes.

But it is the last—or at least the most recent—out-of-Africa episode that, paradoxically, threatens to tear the Okavango apart. Our species, *Homo sapiens*, also evolved in Africa a mere 125,000 years or so ago. Some of us left Africa about 90,000 to 100,000 years ago. A lot happened along the way, and the collision between the newly returned prodigal sons and daughters and their ancestral homeland does not bode well for Eden. But the conflict between the modern world and the ancient setting of our birth does dramatize nicely the very different way we perceive—and fit into—the natural world now, as opposed to the way we were when some of us first left some 100,000 years ago.

SURVIVAL OF THE BIG HAIRIES Our genes say
all of us, all of the nearly 6 billion *Homo sapiens*, originally came from Africa. South African fossils over 100,000 years old are the earliest specimens of anatomically modern humans yet discovered, but perhaps the most suggestive evidence that Africa truly is our ancestral home is simply the survival of the big hairies, or, in more formal terms, the persistence of the African megafauna. Why have lions, hyenas, elephants, rhinos, wild dogs, giraffes, buffalo, so many species of antelope, and so on survived in

such numbers (until very recently, that is), while most of the equally impressive large creatures in all the other Ice Age ecosystems around the world have pretty much disappeared? Why are all the big hairies dead except *African* big hairies? The answer is that we humans evolved in concert with, and literally as part of, the African ecosystems. The big hairies know us well. They and we grew up together, so to speak, in a dynamic equilibrium, an equilibrium still palpable today.

Western safari-goers know that guides on the African savannas will drive right up to a pair of snarling, rutting lions, park right under a leopard snoozing in a tree, and even approach a small herd of old bull African buffalo—the meanest sons-of-guns on the plains. A mild bemused glance is all a Land Rover packed with tourists gets—unless someone stands up abruptly, or, worse yet, hops out of the car. In Botswana, the distinction is all the more dramatic as tourists are still routinely escorted on foot, the guides armed with nothing more formidable than a large knife or perhaps a single flare (guns are not allowed in the Moremi Game Reserve, the large government preserve in the Okavango Delta). As soon as a herd of red lechwe, zebra, or literally anything else—save perhaps the ornery buffalo—spot the upright profile of even a single human being, they melt away imperceptibly if the humans are distant, breaking into a trot if humans suddenly appear close. The big hairies keep their distance: They have been hunted by humans on foot for hundreds of thousands of years, probably for well over a million years.

When presented with the rapidly waning opportunity, the San still hunt on foot, but people in cars are not perceived as hunters, that is, not as the same creature that stalks them on foot. Thus far, in most of Africa, cars pose no threat of predation, and engine noise and the stink of petrol, helping to mask human presence in the first place, rapidly become tolerated. You can drive right up to a big hairy, but you can't get within a half mile of any of them on foot.

Animals have to learn about human predation—as witness the still amazing trust or, actually, insouciance—displayed by all the birds and reptiles of the Galapagos Islands off the Ecuadorian coast. Only recent invaders from the mainland, such as the vermilion flycatcher, show the kind of wariness we expect as normal animal behavior. Now consider what it must have been like when, starting some 100,000 years ago, our species

Homo sapiens began its diaspora, trickling out of Africa and beginning its spread around the globe.

Modern humans arrived in Europe about 38,000 years ago. Within a few thousand years, Neanderthals—considered a separate species by most paleoanthropologists—disappeared, presumably victims of our arrival. We either killed them off or simply outcompeted them in the hominid hunting-gathering niche.

In the other direction, modern humans also reached Australia about 40,000 years ago, triggering a die-off of the larger native species of Australian mammals and lizards. Everywhere you look—especially in the outer reaches beyond Eurasia (where earlier pre-*Homo sapiens* species had lived)—the same pattern appears over and over again. As soon as modern humans arrive, there is a quick die-off, especially of the larger mammals and birds. In the New World, most of the large Ice Age mammals survived to the point when, just a little over 12,000 years ago, humans first crossed the Bering Land Bridge and rapidly spread throughout both North and South America. Immediately, the Pleistocene big hairies—the woolly mammoth, the mastodon, the giant bison, the woolly rhinoceros—became extinct. Eight thousand years ago, the islands in the Caribbean received their first human invaders, and right away many species, especially the larger mammalian species, became extinct. Two thousand years ago, it was Madagascar's turn, and its hippos and giant elephant birds quickly succumbed.

We modern humans were clearly like bulls in a china shop, disrupting ecosystems wherever we went, especially systems that had never had any of the earlier hominid species living in them. Although it is true that this disruption was deeper than simply overhunting (many smaller species also fell victim to extinction at the same time), there can be no question that the larger game animals—the big hairies—were totally unwary. They did not have that aversion to our two-legged profile that seems to be part and parcel of every single large African mammal's take on life. They were slaughtered in large numbers—and anyone who doubts that extinction can come from overhunting need think only of the passenger pigeon, whose numbers were reduced from the tens of millions to absolute zero in just a few decades. Consider the very near miss of one of America's few surviving big hairies, the American buffalo (bison). The native Americans, descendants of the earliest human migration to the New World, had reached an equilibrium with the huge herds of buffalo, an equilibrium quickly disrupted by the advent of Europeans with firearms. Had it not been for public-minded citizens (including Theodore Roosevelt) organized through the New York Zoological Society ("Bronx Zoo"), this magnificent species would have been completely extirpated by World War I.

Back to Africa. If the survival of Africa's megafauna is elegant testimony to the simple fact of that we came from there, its rapidly diminishing numbers are the equally stark result of our return there—return, that is, of prodigal elements no longer leading the life of the hunter-gatherer. The advent of agriculture some 10,000 years ago—the conversion of forest and prairie to controlled monocultural growth, the domestication of livestock—effectively removed humans from their primordial position inside local ecosystems. Overnight, the local ecosystem was in effect transformed into alien territory: The tangled web of plants became a morass of weeds, the animal life at best an irrelevance and at worst a competitor for use of the land or a marauder of livestock.

We newly agriculturized, even civilized, humans return, colliding with the ancient ecosystems of our primordial home. Nor is it just Western-world, European-derived humans who have invaded the southern reaches of Africa: Bantu tribes from the north, including the five tribes of Botswanans, and the tribes of the Zulu nation, the !Xhosans (Nelson Mandela's tribe), and many others are also recent arrivals, coming down from the north not too many years before Europeans began to arrive. These peoples were pastoralists, cattle raisers, and often subsistence farmers. They were no more like the native San and Khoi (Bushmen and Hottentots) than the Europeans.

Nonetheless, surely the arrival of highly mechanized Western-culture produces the greatest strain on and the greatest threat to African ecosystems in general and the Edenlike world of the Okavango in particular. It is hunting and ecotourism, to be sure. It is also the building of roads, towns, and airstrips. It is the control (diversion) of water. But above all, it is the raising of crops and, in northern Botswana, the husbandry of livestock—especially cattle brought by Africa's returning prodigal sons and daughters—that promises to erase even the last vestiges of our ancient homeland.

And as the ecosystem itself is threatened, so too are the humans, who, until very recently, have lived as parts of that system, the San of the Kalahari and Okavango.

THE SAN The story goes that the last license to legally hunt—kill-- a bushman was issued in the early 1950s by the then British colonial government of the Bechuanaland Protectorate. Later, the Central Kalahari Game Reserve was established—designed, with the best of intentions, to keep the local San supplied with game—but in effect, creating a huge, fenced-in human zoo. The experiment failed, as the San dropped their nomadic ways and took on the mores and material culture of the modern world. The San have not fared well in the face of modern invasions of their land.

The San, by tradition, are seminomadic hunter-gatherers. Their bands, like nearly all other hunting-gathering groups the world over, seldom number more than 40 men, women, and children—usually the most that any local ecosystem can sustain. For hunter-gatherers, whether of this era or of millions of years ago, are full-fledged members of the ecosystems in which they live: They rely entirely on their own abilities to wrest a living from the natural productivity of the land. In the case of the San, this has meant that the men hunt wildebeest, giraffe, and other large game mammals, while the women gather tubers and other edible plants, all the while tending the children and the home fires. The entire band "owns" the territory in which they live, and the society is remarkably egalitarian, when, that is, things are working normally.

San troubles began as soon as invaders from the north (black Bantu Africans) and the south (white Europeans) started moving in on their territory, bringing their cattle and other needs for San land: hunting, crop raising, diamond prospecting, settlement areas. The San viewed the animals on the land as theirs and had no compunction about taking Bantu and European cattle—the main touchstone of a range war that quickly led to the San's status in both black and white communities as subhuman, and as such to be exterminated at the least provocation. Hence, the hunting licenses.

Anyone who has seen the dramatic motion picture *The Gods Must Be Crazy* knows that the old Bushmen ways are basically gone. Other films—

such as anthropologist Richard Lee's documentary of the western Kalahari !Kung Bushmen—depict a time almost completely gone. It shows the men making poisoned arrows, stalking a giraffe, finally wounding it mortally, tracking it, cutting it up, and bringing it back for all to share. It also shows the women instructing the children, and finding, excavating, and preparing ground melons and other wild edible plants. Cultural extinction often precedes the actual physical extinction of a people. All too often, indigenous human groups—*especially* hunter-gatherers—are victims every bit as much as are the animals and plants of a ravaged ecosystem.

The very fabric of San society unravels upon contact with cattle-raising agriculturists. When old seminomadic ways are dropped in favor of a sedentary life, traditional egalitarianism instantly vanishes. The men, no longer hunters, do not, as a rule, shift over to lend a hand in the traditional women's arena. Rather, they tend to remain idle. Concept of ownership (so cleverly symbolized by the empty Coke bottle in *The Gods Must Be Crazy*) creeps in, and San society becomes stratified like virtually all postagricultural social systems.

Sometimes it is not so much loss of land as loss of wildlife that does in San societies. In Botswana, the biggest single blow to the San came when, in a rapid spurt beginning in the mid-1950s, two-thirds of the wildebeests, hartebeests, and other large game mammals of the Kalahari lost their lives—an ecological disaster with profound repercussions for the San. This is the heart of today's Kalahari-Okavango story: The rise of the European-supported cattle industry, the immediate disaster for the Kalahari-based wildlife, and consequent virtual doom for the Kalahari Bushmen, as well as the ongoing, very serious threat to the future of the complex of Okavango ecosystems.

THE CATTLE STORY Botswanan beef is delicious, and therein lies a huge, if not particularly delicious, irony. Viewed by some as the lifeblood of the Botswanan economy (ranking just after diamonds as the major source of international exchange), the cattle industry has already destroyed vast stretches of inappropriately designated rangeland. The green expanses of the Okavango beckon to both ranchers and more modest herdsmen alike; the former to expand their domain, the latter simply to take advantage of what they see as otherwise useless land. The Okavango is the one last great refuge for Botswana's wildlife, and if present trends continue, its days may well be numbered.

What killed 2 million wildebeest and other game in the last 40 years? Examine any spot in the Okavango where large herds of antelope and zebra have been grazing and you will find patches of neatly cropped stems,

[Figure 10] *A section of the "Buffalo fence" from the air. The African buffalo stand in the grasses of the Delta. To the left of the fence, the grasses have been overgrazed by cattle and other livestock. The fences were initially built to keep buffalo away from cattle, but pressure is mounting to allow cattle access to the lush grasses of the Delta. Habitat destruction through overgrazing would surely result, leading to the demise of buffalo herds and most other wildlife.*

perhaps 5 centimeters long, forming swards with a distinctly mowed look. The grazing adaptations of all these species—including the African buffalo, the closest thing to a native African cow—are neatly suited to the physiological properties of the various native grasses: Grazing simply does not go too far. No antelope cuts a grass stem down by its roots—much less *including* the roots—as cattle, and especially goats, are wont to do. [Figure 10] The result: Each year, with the returning rains, the grasses green, sending up new shoots, and the cycle continues.

Until, that is, cattle got their teeth into the fragile grasses of the Kalahari. Cattle are not primordially of the tropics; their ancestral homeland lies in the northern temperate realm, and their original fodder were different grass species accustomed to greater dosages of rainfall, rendering them, in a very real sense, hardier than the grasses of the Kalahari. It is the essential *dryness* of the Kalahari that renders the land particularly vulnerable to the excesses of cattle grazing. Overgrazing anywhere else might mean that grasses take longer to recover. But a single bout of overgrazing in the Kalahari, especially if the rains are delayed (or, as has been happening increasingly in recent years, if the rains fail to come), can turn the grassland ecosystem into a dust bowl virtually overnight.

Overgrazing ruined quite a bit of the Kalahari, and the wildebeests and other big vegetarian hairies simply tended to lose out in the competition for grazing room. Not that it is the cattle alone who are their competitors: I've seen films showing native cattle herders literally stoning to death wildebeests that had the bad fortune to get too close to a herd of cattle on their way to that other precious commodity for which both groups compete: water.

Direct competition with cattle for grazing land and water and consequent overgrazing and cattle-caused desertification, as serious as they are, are by no means the whole story behind the dramatic loss of over 2 million of Botswana's Kalahari wildebeest, hartebeest, and zebra in the last 40 years. The nearly 3,000 kilometers of fencing—erected to keep cattle away from antelope and, especially, African buffalo—have had a lot—the lion's share—to do with the sudden demise of these huge herds.

Mark and Delia Owens, two young, idealistic, and (by their own description) naive biology graduate students arrived in the northern Kalahari in 1974, searching for fresh problems in wildlife biology. First considering, but quickly rejecting, the Makgadikgadi Pans, they turned around and entered the Central Kalahari Game Reserve (the "human zoo"), selecting a campsite in Deception Valley, a grassland area within a now permanently dry river bed. As fortunate as they were diligent, the Owens became the first biologists to study in any great detail the habits of the elusive, nocturnal brown hyena, a smaller relative of the far better known spotted hyena (of "laughing" fame). The Owenses observed a good deal more than their brown hyenas, getting to know the local lions and all the other elements of the ecosystem quite well. They recounted their experiences in the compelling book *Cry of the Kalahari*, each one taking turns writing individual chapters in a truly novel and personal story.

It was Mark and Delia Owens who first blew the whistle on the inadvertent slaughter-by-cattle fence of the wild game of the Kalahari. In their last chapter, they recount their experiences along one fence in particular, where they describe dead and dying wildebeest and other antelope festooned in, on, and around the fence in great numbers for long distances. The fences were simply in the way: As all who have watched African wildlife films know quite well, blue wildebeest and their nearly constant companions, Burchell's zebra, are migratory species, moving with the rains in perpetual search of fresh grasslands and open water.

The fence along which Mark and Delia Owens found such large concentrations of dead and dying wildebeests was the Kuki, the northern boundary of the Central Kalahari Game Reserve. In dry times, wildebeest (and originally, zebra, which disappeared from the central Kalahari in the 1960s, victims of a prolonged drought) would migrate north to Lake Ngami, a sort of overflow outpost to the southwest of the Okavango Delta proper. Now the fence was in their way and thousands of wildebeest simply died. Eventually, remnants of the herd, heading east along the fence, managed to find Lake Xau, a far less reliable source of water, near the Makgadikgadi Pans.

Ironically, the Kuki fence actually enhanced the contact between cattle and wild game, at least in the Makgadikgadi region. Even with the saving waters of Lake Xau, the already overgrazed grasslands provided far less sustenance than the expected green grasses around permanent water supplies that wildlife, since time immemorial, automatically seeks out. Nor was the effect of the fence simply to thwart wildebeest moving northward toward permanent water during the dry season: During the *wet* season, many species, including buffalo and elephant, would roam southward out of the delta, spreading out over the terrain that, temporarily at least, had plenty to sustain them.

But it is an ill wind that blows no good. Now that the vastly depleted wildebeest herds have more or less adjusted to the Kuki and other cattle fences, now that the issue is, "How do we save what's left?", we see the other side of the coin. The fences were erected to protect cattle, but now they are actually saving the wildlife, especially that of the Okavango Delta's ecosystems.

Even more notorious than the Kuki fence is the so-called Buffalo fence, erected beginning in 1982 around part of the eastern and all of the southern border of the delta. Leaving the airfield at Maun to fly north and (usually) somewhat eastward into the delta, you soon reach the Buffalo fence, so named as its original purpose was to prevent buffalo from ranging southward out of the delta during the wet season and thus coming into contact with cattle herds. Even in the dry season, there is a stark contrast between the grasslands on either side of the fence. All too often, on the southern side, the ground is bare with nothing but light gray, clayey dust exposed, but on the other side, there is grass—yellowish, perhaps, in dry times, but lushly green when the floods come from the north or when it rains in the delta itself. The Buffalo fence is literally saving the delta, with envious cattlemen, modest herdsmen, and big ranchers alike looking longingly to its other, deltaic side.

Why the fence in the first place? True, cattle ranchers wanted to preserve grazing for cattle, eliminating competitive grass-chewing buffalo and antelope, but the fence keeps cattle *out* of prime delta grasslands. The reason for the fence was the age-old problem of disease, especially nagana, the cattle version of African sleeping sickness. The tsetse fly, for very good reason, is now championed as "the savior of Africa," and, if this is going a bit too far, it is not hyperbole to narrow the focus and apply the slogan to the Okavango Delta, where the tsetse fly, and the particularly virulent form of sleeping sickness that it transmits to cattle and humans alike, thrives. Nagana is invariably fatal in cattle. It was tsetse flies who built the buffalo fence. [Figure 11]

Antelope and other game are also bitten by tsetse flies, and their blood harbors the microbe (a trypanosome) that, in humans and nonnative cattle, causes the fatal disease. Antelope themselves are seldom diseased, but

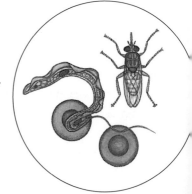

[Figure 11] *A tsetse fly, plus an enlarged view of a try-panosome attacking red blood cells in a mammal that the fly has bitten. The bite of a tsetse fly is surprisingly painful, and the flies are incredibly persistent. We were once attacked by a swarm of these flies near sundown as we drove along a small Okavango stream. Our driver sped up, but the flies hung on despite the winds whipping through the car's cabin. Fortunately, no one came down with sleeping sickness!*

are the Typhoid Marys of sleeping sickness/nagana. An early remedy was simply to slaughter whole herds of antelopes in an effort to diminish the pathogenic reservoir. Far more humane, then, to have a buffalo fence and to pursue a campaign of tsetse fly control and eradication (most visible to travelers in the delta as small blue tents housing pheromonal attractants, luring the flies to their deaths).

Sleeping sickness/nagana is not the only disease prompting the erection of all these thousands of kilometers of fencing. Perhaps equal to the fear of cattle contracting nagana from the bites of tsetse flies is the concern that antelope and buffalo, all members of the cattle family, may pass the virus that causes hoof-and-mouth disease directly to cattle. The EEC (European Economic Community), the major international market for Botswanan beef, and in effect successor to the World Bank as the major subsidizer of the Botswanan cattle industry, insists on strict control of hoof-and-mouth so as not to expose *European* cattle to outbreaks of the disease. Most recently, new fencing near Botswana's border with Namibia (to the west of the delta) is going up to reinforce the *existing* fence at the border proper—all to stem the inflow of cattle from Namibia, cattle that are bringing with them a deadly lung disease. Here, the enemy is other cattle, not wildlife, but the fences will, of course, have their effect on the wildlife as well.

Now that the fences have exacted their central Kalahari death toll, now that the remaining herds of wildebeest and other game species seem to have adjusted to the realities of the fence, the shoe is on the other foot. When, in 1995, a rumor started in the delta to the effect that there was now "scientific evidence" that buffalo do not in fact transmit hoof-and-mouth disease to cattle (a rumor whose truth I could not substantiate), it was widely regarded as a myth, even a hoax invented to further the cattlemen's desire to expand into those lush, green, inviting grasslands of the delta. They would rather take their chances with nagana and hoof-and-mouth disease and have those grasslands, as the industry continues to expand, and cattle continue to die from the vicious one-two punch of drought: no water to drink; no grass to eat.

Small wonder, then, that many conservationists in the delta want to cut their losses and change their initial opposition and support at least the fences immediately ringing the delta. The damage has been done, and now the tsetse fly-buffalo fence combination seems the last best chance for the Okavango Delta's ecosystems to survive. Even now, cattle are encroaching deep into the delta, especially up north, in the "panhandle," where there are no fences and no national parklands. Where you see cows, you don't see red lechwe, let alone buffalo and elephant.

OTHER PROBLEMS Cattle are the major story, the major manifestation of the impact of the hand of man, particularly agricultural practice, on natural primordial ecosystems. But they are by no means the entire story. Plans to divert Okavango waters—at first for cattle use, more recently to supply governmental buildings in Maun—have met with outrage from the local tribal council and others. Perhaps a more serious threat is diversion upstream, a familiar theme the world over, as Namibia and Angola each have announced plans to divert some of the water that otherwise ends up in the Okavango Delta.

There is, in addition, fire, a natural phenomenon now understood to be essential to continual grasslands ecosystem renewal. Southern Africa has more lightning strikes per year and more human deaths from lightning than any other part of the world. Still, humans are setting fires in record numbers in the delta, and suspicion is growing that too many fires will debilitate, not strengthen, the grasslands, mimicking the effects of overgrazing by domestic livestock.

Other difficulties are harder to assess—at least the extent to which humans are involved. Take, for example, the current drought. Droughts are natural, of course, but the current debate on the effect of global, collective humanity on the planetary system, particularly world climate, is at least arguably relevant to understanding current changes in weather patterns in southern Africa. Even with the drought, we may suspect some contribution of *Homo sapiens*.

Or consider what many take to be the recent population explosion of elephants in the delta. Among the most familiar and best loved of all the big hairies, "ellys," as delta residents still fondly call them, are profound modifiers of their environment. Elephants routinely enter camps at night, tearing down trees right and left—mostly to secure ripe fruit or favored leaves, but apparently in other instances, to create a bit of mischief as well.

Tree damage is markedly on the rise in the Okavango, and no one wants to see a repeat of the East African Tsavo story, where elephants literally trashed that reserve during a prolonged drought, when too many elephants had too little to eat. The problem is worse to the delta's east, along the Chobe and Linyanti, but there are ominous signs that elephant damage will increase in the delta itself.

Elephants were not always common constituents of the delta ecosystem, a fact that is hard to understand because they seem to be doing so well there now. Why this upsurge in elephant population in northern Botswana? As usual, the causes are a complex amalgam, including the current drought. Surely fencing impeding their migrations south out of the delta and, most particularly, wars (especially in Angola), hunting, and increasing conversion of natural elephant habitat into agricultural fields and grazing lands in Angola, Zambia, and Zimbabwe are contributing factors. Elephants are retreating to the delta as their more usual haunts are taken away. They are doing well there, and their numbers are increasing to a point where, perhaps soon, something will have to be done to control their numbers. It is a shame when too much success means that humans, once again, feel compelled to manipulate and regulate a system.

Even ecotourism, so often touted as a major hope for saving ecosystems such as the Okavango Delta, poses real problems. With tourists come speed boats, airplanes, helicopters, four-wheel-drive vehicles—not to mention water usage, disposal of human and other solid wastes, clearings for buildings and airstrips, and so on. More subtly, no human can observe a system—*any* system—without disturbing it in some way. When a San hunter-gatherer walked out on the plains to hunt, he was observing, and in a very literal sense disturbing the system, though in ways he sought to minimize and in ways that were very much like interactions of other components of that system. The San were part of the system.

Ecotourists are not. As we shall see in more detail in the following chapter, no postagricultural, let alone postindustrial, people are part of any local ecosystem whatsoever, not even the systems in which each of us lives. Because I do not derive my food, clothing, or shelter from it, I am not part of the (highly modified) local ecosystem in the woodlot in my backyard. Still less can a tourist from North America be construed as a part of the Okavango system. We are strictly observers, perhaps prudent ones, but our presence, the mere act of looking at the wildlife, has its effect on the system. The effects range from the obvious, egregious low flyovers that start buffalo herds stampeding, to the destruction of grasslands as big hairies are pursued in four-wheel-drive vehicles over the savannas, to the much more subtle effects of simply being noticed—as when a small band of serious questers disturbs a browsing bull elephant as they photograph him.

WHAT'S GOING TO HAPPEN? The Okavango
is unique, a fantastic remnant of our own ancestral climes. The same kind of story of threat, even imminent collapse, can be told in absolutely every ecosystem in the world. Human encroachment has reached to every conceivable ecosystem, and many have already been obliterated. What makes the Okavango situation so poignant, so especially compelling, is that it is a still-healthy vestige of our own heritage.

What can be done to save the Okavango? For if left untended, if in other words humans do not regulate themselves, the Okavango Delta is surely doomed. The fences are a good, if inadvertent, start. But fencing in the long run never works: Good intentions—more often than not stemming from well-meaning, conservation-minded folks from far away—can have no lasting effect unless the economic needs of indigenous peoples are met.

It is already too late for most of the San, whose culture has already largely been lost. Most San today live in villages and are integrated to varying degrees into the contemporary Botswanan economy. They and members of the Bantu-derived Botswanan peoples continue to eke out a living as modest agriculturists and tenders of small herds of cattle. It is notorious that wildlife cannot be saved unless habitats—intact ecosystems—are set aside for wildlife, yet this is impossible unless indigenous peoples can come to view the wildlife as economically beneficial. People living on the margins of the delta naturally gaze across the fence with envy at its waters and grasslands. They see the wildlife as competitors.

"Don't shoot the poacher, hire him" is a refrain that mirrors the growing realization that, without involving indigenous peoples directly in the conservation effort, all will be lost. Stronger yet is the movement (called the "Campfire Movement" in East Africa) to invest local people with literal ownership of their local wildlife. There is no hope for the long-term survival of wild places like the Okavango until wildlife is viewed as a renewable resource—a valuable resource for food, for granting hunting permits, and most especially for the explosion of interest in ecotourism—and until local peoples can participate and profit from the existence of wildlife.

The Okavango belongs to all of us in a real, if general, sense, simply because it mostly closely resembles the place where all of us were collectively born. But its fate lies in the hands of its current human caretakers, those who live there now and who must do the real work if the Okavango is to survive. Only if these people—black, white and San—can share more fully in the fruits of a still-pristine Okavango Delta does this last vestige of Eden stand a chance.

Chapter 2 BIODIVER- SITY, EVOLUTION, AND ECOLOGY

Imag-

ine looking out a window at the living world. Even in the in-

ner cities there are species other than Homo sapiens *to be*

seen. Though many are the familiar commensals of human

settlement, such as rock doves (pigeons), sparrows (most often

the European house sparrow), (European) starlings, and Nor-

way rats, there are also domesticated dogs and cats, and ves-

tiges of pre-city ecosystems (squirrels come to mind). The root

of all ecosystems is, of course, photosynthesizing plants, and

every city still features a wide assortment of them, from invad-

ing weeds to carefully cultivated trees, from exotic species to

remnants of the local natural ecosystem. The native Ailanthus

tree is the famous "tree that grows in Brooklyn."

Life occurs virtually everywhere on the globe, and everywhere you find life, you find more than just one distinct form, more than just one species. There are photosynthesizing plants that form the base of the food chain of all local ecosystems (all save the peculiar vent faunas of the abyss, which are fueled instead by bacteria that metabolize chemical compounds formed in the heat emanating from fissures in the oceanic crust). There are microbes that cycle nutrients, contribute to the decay process, and supply vital elements and nutrients to the "food chain," the flow of energy from the "primary producers" (the photosynthesizers), up through the animals and microbes that eat plants, and eventually through the often intricate chain of animals that eat the animals that eat the plants.

This is one distinct facet of life: The local ecosystems, where members of different species live cheek by jowl in a complex fashion, systems where energy and nutrients are abstracted from the physical environment and cycled through complex networks of interdependency. That's what you glimpse as you look out your window: Some of the components of the local ecosystem that is up and running virtually everywhere on Earth—even, however distorted the form, in the world's inner cities.

There is another, equally distinctive side of life, one that can also be detected through your window. It is the evolutionary world of species. Here the connectedness is not between populations of different species in one local spot, but rather the connections between, for example, the squirrels in your backyard with those living across the street, in the next town—and indeed (in the case of the eastern gray squirrel, *Sciurus carolinensis*), up and down the entire eastern seaboard of North America, extending as far west as the Rocky Mountains. Energy does not flow through, holding a species together—as it holds together local ecosystems. Rather, *genetic* connections unite species. Gray squirrels in Atlanta share the vast majority of their genes with those living in New York and Toronto (not to suggest that squirrels are purely urban these days!). They share, as well, complex mating behaviors that would enable a southern squirrel to recognize a Canadian eastern gray squirrel as a potential mate. Mating and sharing a common supply of genes holds species together.

But there is more to the evolutionary side of life than simple commonality of gene pools—the existence of separate species. As everyone recognizes, there is more than one kind of squirrel living in North America. On the other side of the Rockies, the eastern gray squirrel is replaced by the closely similar *western* gray squirrel (*Sciurus griseus*), a species that nonetheless differs from its eastern counterpart in a number of minor, but consistent, ways. Most critically, the two species do not share a common mating behavior and do not recognize each other as potential mates even when they do come in contact.

Geography plays an important role in the evolutionary story. A species is seldom confined to a single, localized area. A species' members are typically spread out over wide areas, and its local members are often playing economic roles in a number of somewhat different kinds of local ecosystems. Gray squirrels are equally at home in parks, suburban backyards, and the deep mixed-hardwood forests of eastern North America. As we have just seen, closely related species are often found living in separate, if closely adjacent, regions—a clue that tells us much about how new species evolve from old.

But a glance out the window might also reveal more than one kind of squirrel living in the same local spot. Around New York City, eastern chipmunks (*Tamias striatus*) live alongside gray squirrels. Chipmunks may be brown, with white and gray stripes, but most of us have no trouble seeing them as some kind of a squirrel. Their faces and behaviors seem similar to those of other squirrels, with the most conspicuous difference perhaps coming in where you normally see chipmunks—on the ground. Gray squirrels forage on the ground but take off into trees when alarmed; chipmunks usually run for rock piles. Eastern gray squirrels nest in trees; chipmunks mate and nest in nooks and crannies on the ground, seldom venturing more than a few feet up the trunk of a tree.

Chipmunks are ground squirrels; eastern grays are tree squirrels. They prefer somewhat different food items and utilize different resources for shelter. They are ecologically sufficiently unlike, sufficiently noncompetitive for resources, that local environments typically can support populations of both species. Put another way, local ecosystems are composed of populations of species, each of which plays a discernibly different role in the cycling of nutrients and the flow of energy within the system.

We have not quite exhausted the list of squirrels that live in eastern North America and are visible through at least some suburban and rural windows. Woodchucks ("groundhogs," *Marmota monax*) are also squirrels.

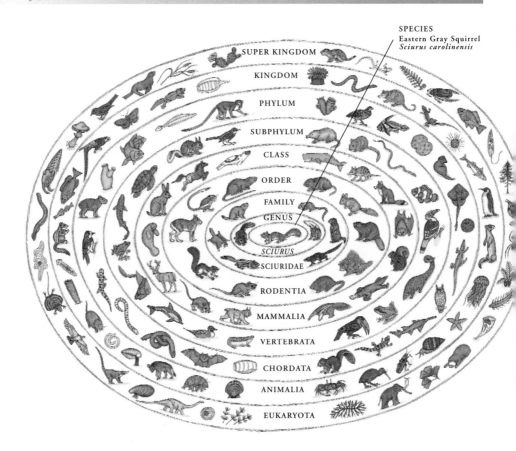

SPECIES
Eastern Gray Squirrel
Sciurus carolinensis

SUPER KINGDOM

KINGDOM

PHYLUM

SUBPHYLUM

CLASS

ORDER

FAMILY

GENUS

SCIURUS

SCIURIDAE

RODENTIA

MAMMALIA

VERTEBRATA

CHORDATA

ANIMALIA

EUKARYOTA

[Figure 12] *An eastern gray squirrel's eye view of the evolutionary interrelationships of all life. The squirrel sits within a set of rings, each progressively including the next larger set of its evolutionary relatives: closest squirrel kin (Genus Sciurus), then other kinds of squir-rels (Family Sciuridae), then other rodents (Order Rodentia), then other mammals (Class Mammalia), etc.—including all of life (except bacteria) in the outermost ring. Precisely the same sort of diagram could be constructed for each of the 10-13 million species of or-ganisms currently on Earth—as well as for each of the countless millions of species that have evolved and become extinct in the past 3.5 billion year history of life. All species, liv-ing and extinct, are related through evolutionary descent with modification.*

These large, ungainly rodents, which lack the long tail and lithe bodies of tree squirrels and chipmunks, stretch the limits of casual impressions of what squirrels *are*. Yet it is true enough: Sharing a number of features, such as tooth number and shape, and details of skull and muscle anatomy, woodchucks (and the four marmot species of western North America) are indeed squirrels as well—albeit fat, specialized, ground-living squirrels.

That is the evolutionary chain. Every organism you can spot from your window belongs to a species, a set of organisms sharing many of its same genes and mating proclivities. Virtually every species has closely similar related species, typically living in adjacent but separate regions. Each species also has more remotely related species with which it shares a more general similarity, and these different species might be found living in the same local region. Each species has other, even more remotely related species, species that it no longer closely resembles. An ever-widening circle of evolutionary relatedness embraces every species on Earth. Ultimately one finds the similarities, the shared features that are the surefire telltale signs that absolutely every species on Earth is related to absolutely every other species on Earth in a pattern of evolutionary connectedness every bit as complex as the patterns of energy flow that link all the disparate populations of different species within the confines of a single local ecosystem.

Thus, groundhogs are squirrels, but squirrels are rodents—a huge group of species, all of which share a number of unique features—most especially front teeth (incisors) which are ever-growing. Beavers are also rodents, as are rats, mice, voles, lemmings, and many other groups as well. Rodents are mammals, united by the possession of hair, three middle ear bones, and mammary glands. And mammals are amniote vertebrates, sharing the amniote egg with reptiles and birds. All these groups are vertebrates, and vertebrates are but one division of animals. Animals share a common cellular anatomy with plants, fungi, and some microbes: the so-called eukaryotic cell with a separate nucleus housing the bulk of the cell's genetic machinery. Finally, absolutely all living species share one feature in common: All possess the molecule of heredity known as RNA, ribonucleic acid (most also have DNA, deoxyribonucleic acid). Thus, bacteria and archaebacteria—the prokaryotes—are linked with all the eukaryotes to form the unified tree of life that we will explore extensively in the next chapter.

Early naturalists saw these nested patterns of similarity linking Earth's species. The Swedish scholar Karl von Linné (whose latinized name, Carolus Linnaeus, is perhaps more familiar) saw that each species is connected by similarity to others with which it forms a basic group that he termed the *genus*. Think of *genus* as "general" and *species* as "specific," and Linnaeus' basic approach becomes clear. The eastern gray squirrel is *Sciurus carolinensis*, and its western counterpart is *Sciurus griseus*. Two different species of the same genus. Fox squirrels, though more distantly related, are also conventionally included within the genus *Sciurus*, but red squirrels are sufficiently remotely related that they have their own genus, *Tamiasciurus*, as do all species of chipmunks (*Tamias*). Groundhogs (genus *Marmota*) are so different that they form a whole subgroup. Squirrels as a whole are an entire Family of the Order Rodentia: Family Sciuridae. Figure 12 shows the classification of life from the vantage point of the eastern gray squirrel, *Sciurus carolinensis*.

It was Charles Darwin who convinced the world that the nested pattern of resemblance linking all species and providing the basis for Linnaeus' hierarchical classification scheme has only one plausible explanation: All living species are descended from a single common ancestor in a process Darwin termed "descent with modification," and which we have come to call, more simply, "evolution." All living species have inherited vestiges of that original single-celled microbe, the RNA found in absolutely every living species. As time goes on, as lineages diverge and new features are developed, they are passed on to succeeding generations and species but not found in separate lineages, which develop their own peculiar attributes. The more closely related any two species are, the more genetic information they will share, and the more similar they will appear. The squirrel story is repeated starting with absolutely every other species on Earth: from the simplest of microbes to ourselves, *Homo sapiens*.

THE ECOLOGICAL HIERARCHY The evolutionary chain is nicely matched by an equivalent ecological chain of interconnectedness. On the evolutionary side of the equation, genes held in common link up members of a single species and closely related species into an ever-widening array of life forms connected through common evolutionary descent. In ecological systems, the flow of energy on a moment-by-moment basis, rather than shared genetic information, provides the linkage, the internal cohesion that defines local ecosystems and connects those local systems regionally and ultimately the entire world over.

Every living organism requires energy to develop from a fertilized egg or seed (in the case of complex organisms such as plants and animals) or to divide and continue to exist in the case of microbes such as bacteria and protoctistans (single-celled eukaryotic organisms). Organisms need energy to build their bodily structures—tissues, organ systems—and to maintain these structures throughout their lifetimes. They need energy to reproduce: In the case of animals, to produce eggs and sperm, to mate, and, in the case of mammals, to bear live young and rear them. Life is a constant search for, and consumption and utilization of, energy.

Animals and some microbes are *heterotrophs,* meaning that they obtain their energy requirements from the consumption of other organisms, other animals, plants, fungi (mushrooms, yeasts), and microbes. Plants and some microbes, in contrast, are *autotrophs:* They synthesize their food directly by literally "eating" sunlight, converting solar energy to stored energy in the form of sugars, produced through photosynthesis. Fungi, the indispensable agents of decay in terrestrial ecosystems, are *saprophytic,* meaning that they derive their energy requirements by direct absorption of the tissues of dead plants, animals, microbes, and even other fungi.

An ecosystem is a place, an arena where energy flows constantly from one living component to another. These living components are not species, but rather *local populations* of various different species. It is, in large measure, the local population of squirrels living in a patch of forest, dependent on the acorns produced by oak trees, and being eaten by local red foxes and broad-winged hawks. But the local ecosystem is much more than the organisms in it. It is, in addition, the flowing energy itself, plus the inorganic components—the chemical composition and temperature of the sediment, soil, water, or atmosphere, depending on the nature of the system, be it a lake, a section of prairie or a montane meadowland. It is the nutrients—various organic and inorganic compounds, such as vitamins— that are found locally. Finally, the local ecosystem is a place, a pond, say, or a beach. All these components go into the mix and define the dynamic web of life that *is* a local ecosystem.

Where are the boundaries of local ecosystems? A lake might have a distinct shoreline, but the plants fringing its edges—those in the shallow waters as well as those living down close to the edge—are invariably different from plants (like water lilies) living in deeper waters or the trees and shrubs living on drier ground farther away from the shoreline itself. How do we recognize the boundaries between ecosystems?

Ecosystem boundaries are inherently fuzzy. If we say a pond is an ecosystem, and its surrounding forest is an ecosystem, what do we make of the osprey—a fish-eating hawk—that perches, rests, and nests in forest trees, but plunges feetfirst into the lake to get its food, its all-important energy source? Clearly local ecosystems are connected, and they are connected by energy flowing across their blurred boundaries. The productivity (photosynthesis-based growth of algae and plants) in the pond supports a complex web of life within the pond: Insect larvae, freshwater clams and snails, crustaceans, leeches, and fish, all eat one another, but the plants and algae form the base of the "food chain." That dynamic system is linked irrevocably with its neighbors: the fishing osprey; the beavers that live on the pond but utilize the trees fringing the pond; the streams with decaying organic particles and nutrients flowing into the pond; and the stream that invariably flows *out* of that pond.

Local ecosystems, then, are places where most of the energy is cycled through growth, death, and decay of a particular set of organisms; but energy is always flowing into and out of these systems. If no man is an island, no ecosystem is entirely self-contained. If it is energy that links the members of a local ecosystem, it is also energy that flows between systems, linking them into regional systems, such as the lakes, rivers, and forests of northeastern regions of North America. You can recognize finer and finer subdivisions of these systems—but also broader and broader regional systems—all interlinked through energy flow. Local ecosystems are connected to adjacent ecosystems, forming regional systems that themselves are parts of continentalwide, or oceanic, systems. Ecosystems, like species, are parts of larger-scale systems that are linked in a smaller-within-larger hierarchical fashion.

Ultimately, the entire surface of the globe is linked through energy transference—through organisms, and through such physical phenomena as wind, rainfall, and fluctuating temperature regimes. This is the Biosphere, the global ecosystem. This is the system to which humans still belong, notwithstanding our newly found emancipation from the productivity—the carrying capacity—of purely *local* ecosystems.

Thus, ecological systems are dynamic amalgams of parts of many different species. Ecosystems seethe with energy and are literally the arenas where the pageant of life is enacted. The players in the arena—the local populations of organisms—belong to species that are parts of a 3.5-billion-year history of evolutionary diversification. Thus, biodiversity has two faces: (1) all the species that are related to others, mostly living elsewhere, in a complex hierarchical network of evolutionary kinship; and, at the same time, (2) the rich array of *different* species present locally and interacting in complex ways as energy, nutrients, and matter are extracted from the physical environment and cycled through the living system.

Though evolutionary and ecological systems are fundamentally different in nature, they are nonetheless opposite sides of the same coin: Evolution and ecology go hand-in-hand, linked by processes we shall now explore.

THE ECOLOGICAL CONTEXT OF THE EVOLUTIONARY PROCESS

The heart of the evolutionary process lies squarely inside local ecosystems, where the moment-by-moment, day-to-day problems of the game of life—struggles to find sustenance and shelter from the elements and predators, as well as the seasonal opportunity to reproduce—are played out. Here is where the action is and where the fundamental process of natural selection records, in effect, the winners and losers in the game of life.

Charles Darwin is famous, most of all, for his enunciation of the principle of natural selection. The idea of evolution—that all organisms are descended from a single common ancestor—had been around, in one more-or-less vague form or another, ever since the Greeks. It took the Herculean efforts and keen insights of Charles Robert Darwin to convince the thinking world that evolution must have happened. He did so in his epochal book, *On the Origin of Species by Means of Natural Selection, or the Preservation of Favoured Races in the Struggle for Life,* first published in 1859. The book's title (especially the longer version, seldom mentioned these days—everyone calls the book simply *The Origin*) gives a hint on why Darwin was successful where so many before him had failed to mount a convincing argument that life has evolved. Darwin produced a plausible mechanism for evolution, where no one else before him had. And though it is true that the great naturalist Alfred Russell Wallace dreamed up natural selection at the same time (literally—Wallace first encountered the basic ideas while in the throes of a malarially induced dream), it was Darwin who meticulously gathered the supporting evidence and, ultimately, made the brilliant book-length argument that carried the day and established the idea of evolution as a respectable scientific concept.

Natural selection is a very simple idea. It comes from combining two very different kinds of observations about organisms in the wild. The first is that no two organisms in a population (save identical twins) are exactly alike. Such natural variation is present in every population in every ecosystem. Moreover, this variation is heritable, meaning simply that organisms tend to resemble their parents. Darwin had a wildly inaccurate theory to explain heredity (i.e., how it is that organisms do in fact resemble their parents), but all he really needed to know to formulate the principle of natural selection was simply that organisms *do* inherit their features from their parents.

Darwin had help from an unexpected source in seeing the second key ingredient to his fledgling idea. Sixty-one years before Darwin's *Origin* appeared, the Reverend Thomas Malthus had published a paper entitled "An Essay on the Principle of Population, as It Affects the Future Improvement of Society." Though Malthus confined his analysis to humans, Darwin (and, independently, Wallace) awakened to Malthus's point that, left unchecked, the populations of any species will grow geometrically, mushrooming to huge numbers in a relative brief span of time. Darwin himself calculated that, starting with a single pair of elephants, and assuming a production of six offspring per pair over a 60-year period of fertility, some 19 million elephants from that single original pair would be born in the relatively brief span of 750 years! Something—some set of factors—must be at work keeping each and every population of each and every species at more-or-less constant numbers.

Those factors are precisely the exigencies of the game of life: the need to find energy sources and nutrients, the need to stave off disease, to ward off predators, to withstand the rigors of too much or too little water or temperatures that are too hot or too cold. Of these, the inherent, inevitable competition for energy sources (for animals, simply food) is the primary limiting factor of local population size. An ecosystem can carry only so many organisms of a given species. Each species population in a local ecosystem has its own *niche,* meaning its role and function within that system. The niche requirements of any given species population arise from the anatomical, physiological, and behavioral characteristics of organisms of that species. These genetically based features are evolutionary adaptations developed in the dim recesses of geologic time.

In other words, the size of a local population of any species in a local ecosystem is limited by the availability of critical resources such as food and by the different ways each species is adapted to utilizing those resources. On a given patch of African savanna, there are usually far more impalas than there are, say, sable antelopes—a difference in numbers that reflects the fact that sables are larger and more selective in grass consumption than are the smaller, less finicky impalas.

Before putting together these two ingredients—heritable variation and natural control of population size—to yield the dynamic motor of evolutionary change, natural selection, one more point needs to be made about the Reverend Malthus' ruminations on the tendency of populations to mushroom if something happens to increase the carrying capacity of the environment. It is a point that lies at the very heart of the present biodiversity crisis.

Darwin was diligent, and in his quest to understand how Nature herself might be causing the characteristics of animals, plants, fungi, and microbes to change through time, to *evolve,* he studied how animal breeders can modify the traits of their livestock and pet species. He quickly saw that breeders allow only a relatively small number of individuals that have the desired trait in the best developed form to produce the next generation—the better to develop the trait even further. Their offspring, of course, will tend to inherit the traits of their parents.

Darwin's (and Wallace's) great insight is that the factors controlling the size of populations in the wild act in a way similar to the reproductive control imposed by breeders over their stock. In populations where numbers remain relatively constant year in, year out, and from generation to generation, *only those organisms best suited to making a living, to coping with the very factors that limit population size, will, on average, be the ones that survive and reproduce.* Those that are successful, or relatively more successful, at producing offspring will tend to pass along the traits that were the ingredients for success. In more modern terms, those best suited to the environment will tend to leave more of their genes to the next generation.

Thus, heritable variation plus limits to population size equal differential reproductive success—in other words, natural selection. Natural selection has the status of a scientific law: It *must* be true, given the mere presence of heritable variation and the demonstration that population numbers remain more or less constant through the generations. Even though Darwin was totally ignorant of modern genetics (which had not been invented in his time—though there is said to be an unopened copy of Gregor Mendel's early research paper on the shelf at Darwin's residence, Down House), he was able to pinpoint how useful traits—evolutionary adaptations—might be formed through time.

One more point remains: If the traits for success of the preceding generation are sifted by natural selection and handed down preferentially to each succeeding generation, won't evolution soon come to a halt, as traits are perfected? The short answer to that question is yes—as long as the environment does not change. But the environment does change—fluctuating daily, seasonally, yearly, and certainly over longer periods. As long as the basic variation is there, new traits will evolve to meet the changing demands of the environment. *How* that process works takes us deeper into the inner workings of evolution within the ecological arena.

THE RED QUEEN

The Red Queen in Lewis Carroll's *Through the Looking Glass* complained that she had to run constantly just to stay in one place. In the 1970s, that image appealed to University of Chicago evolutionary biologist Leigh Van Valen, who thought it a useful metaphor for how species relate to one another in the wild and how their interaction affects their evolutionary fates.

Red Queen imagery usually focuses on one species living cheek by jowl with many others. What happens, for example, to a population of a particular species when another population it relies on as part of its food supply suddenly becomes scarce, or yet another population, perhaps a predator, becomes more abundant? The Red Queen imagines that any such changes in any other species population represents a (generally) negative situation for the focal species so that the focal species has to "run hard"—meaning change through natural selection—just to "stay in one place"—meaning simply to continue to exist in that particular local ecosystem.

Imagine, for example, a population of gray squirrels living, say, in New York's Central Park. Park rangers recently introduced barn owls to help stem the tide of rats. To the extent that gray squirrels compete with rats for some of their food items, fewer rats would be a good thing for Central Park's squirrels, but barn owls also take squirrels, which is definitely not good for the squirrels. Complex ecosystems—with many different species eating many others, and with competition for finite resources between species with similar needs—mean that the least little change in one population has implications, even evolutionary consequences, for many others in that same system.

Small wonder that biologists looking at the details of ecosystem dynamics, and year-by-year patterns of evolutionary change within populations, see a seething world of ups and downs, with natural selection constantly changing the average characteristics of many species. Years of research by Princeton University's Peter and Rosemary Grant on the famous finches from the Galapagos Islands (often called "Darwin's finches") have revealed great flux from year to year in bill size and shape, a reflection, in turn, of great fluctuations in numbers of various different shapes, sizes, and hardnesses of various seed species that the finches eat. No doubt about it, natural selection is constantly at work, modifying adaptations to suit current conditions. [Figure 13]

Many biologists, looking at evolution over longer time intervals, have noted that, despite the constant change apparent in most local populations from one generation to the next, species are rarely modified consistently in one direction for enough time so that significant evolutionary change can accumulate. Even the Galapagos finches seem to oscillate back and forth, not really going anywhere in an evolutionary sense.

Evolution is in action, certainly, but the dynamics of change as a response to ecosystem perturbation seem more cyclical than directional and therefore not the stuff of the evolution of truly new species or the development of major new adaptations. It turns out that the vagaries of short-term environmental change themselves tend to be cyclical, making it unlikely that natural selection will keep pushing a species in any one particular direction for too long. In addition, every species is broken up into local populations, and each of these populations belongs to a different ecosystem, differing in details from neighboring systems. Thus, the Red Queen nat-

[Figure 13] *The justly renowned "Darwin's finches" from the Galapagos. Darwin himself did not realize the significance of variation in beak size and shape in the distributions of these closely related species on the various Galapagos Islands until British ornithologist John Gould sorted out the specimens Darwin brought back from his voyage on the HMS Beagle. These much-studied birds have become a classic case of natural selection + isolation leading to evolutionary diversification.*

CACTUS FINCH

SHARP-BEAKED GROUND FINCH

COCOS ISLAND FINCH

WARBLER FINCH

LARGE CACTUS FINCH

LARGE GROUND FINCH

MANGROVE FINCH

MEDIUM GROUND FINCH

WOOD-PECKER FINCH

SMALL GROUND FINCH

SMALL TREE FINCH

VEGETARIAN FINCH

LARGE TREE FINCH

MEDIUM TREE FINCH

ural-selection history of each population within any species will differ from place to place: The mix of predators and food species confronting gray squirrels in Florida is decidedly different from that encountered by squirrels in Ontario, making it a near impossibility that the entire species of gray squirrel will be modified by natural selection to change in any one particular direction as time rolls on.

Something else, it seems, is necessary for larger-scale, more permanent evolutionary change. That something else resides not in some new, undreamed-of genetic mechanism, but rather in larger-scale interplay between the twin faces of diversity—the ecological arena and the populations and species that are the actors in the drama.

ECOSYSTEM DISRUPTION AND EVOLUTIONARY CHANGE

From Darwin on down, it made sense for biologists to see evolution simply as natural selection "tracking" environmental change, whether the "environment" was other species in the ecosystem or aspects of the physical environment itself. The standard presumption has long been that, if the environment remains more or less constant, natural selection will work to hone adaptations, perfecting them still further. If the environment changes, then adaptations will gradually be modified to fit the new conditions.

We have just seen that long-term natural selection seldom, if ever, acts to modify significantly the adaptations of a species gradually through time. Rather, when environments change, the species themselves do the tracking by relocating to regions where the environment—the animal, plant, and microbial communities, plus the climate—falls within the range of accustomed haunts. Thus, even in the face of significant environmental change, adaptations of species will remain stable as long as that species can continue to locate familiar environmental conditions.

Older biologists saw the outcomes of environmental change as either (1) evolution, provided that sufficient genetic variability was present, or (2) extinction, if such genetic variation were not forthcoming. Nowadays, biologists accustomed to monitoring significant shifts in animal and plant species distributions just over the past 100 years realize that the most expected outcome of environmental change is species relocation. If relocation fails, extinction is the next most likely outcome.

If failure to locate a viable environment means extinction, how do new species appear, and how do the new adaptations that fill the pages of evolutionary history actually evolve? Paleontologists now realize that ecosystems—and the species whose populations staff those ecosystems—are remarkably stable entities. University of Rochester paleontologist Carlton Brett and his colleagues have pointed to a succession of some 8 stable Paleozoic communities in eastern and central North America that lasted, on average, about 5 or 6 million years. There are anywhere from a few dozen to several hundred marine invertebrate species known from each interval, and few if any of those species show any appreciable evolutionary change for millions of years.

Moreover, on average, only 20% of the species survive to make it into the next interval of stable ecosystems and stable species. Ecosystem disruptions are the key here, with loss of most species—and new appearance of different species to take their place—at an abrupt boundary. The change is relatively sudden, and severe, and without further disruption, ecosystems and species together settle down for yet another long period of relative quiescence and stability. For example, mammals did not diversify appreciably until after the terrestrial dinosaurs had finally, at long last, succumbed to extinction. Evolution appears to depend in large measure on prior extinction—as if the new species and novel adaptations cannot appear unless and until preexisting species and ecosystems are profoundly disrupted.

What is happening at these boundaries? It is difficult to pinpoint what is changing in the environment of those ancient 400-million-year-old seas and 65-million-year-old terrestrial environments. We can look to more recent, yet very similar examples, and get a much clearer idea of what is going on, what causes the stress, and what the evolutionary and ecological dynamics of these sudden turnovers of living systems are all about.

Recall some of the key points in early human evolution recounted in the previous chapter; they have much to tell us about how the entire evolutionary process works. Five million years ago, a global cold snap struck, the Mediterranean dried up completely, antelopes reached Africa from Asia for the first time, and forests began to give way to grasslands, as a cooler, drier climate was established in eastern and southern Africa. Many species disappeared, and others took their place—including the newly arrived antelopes. Two and a half million years later, a similar event occurred: Average global temperature went down by 10° to 15°C, over a period of perhaps 200,000 years. This time, the open grasslands and savannas reminiscent of today's Serengeti spread down the East African Rift Valley system, and, once again, many of the species living in the relatively wetter and more completely wooded setting suddenly disappeared, quickly replaced by different species. Only a few of the older species hung on to take up residence in the made-over ecosystems. Of those that disappeared, one was our own ancestor *Australopithecus africanus*. Happily, of the new species to take up residence after the shock, two were of our own lineage. One was a species of the specialized herbivore *Paranthropus* and, as such, collateral kin rather than our direct ancestor. The other, the famous *Homo habilis*, was most certainly our ancestor. This event 2.5 million years ago took away one hominid species and replaced it with two, an intriguing hint as to what really is going on when ecosystems are disrupted and species severely affected.

Once again, what is going on has both an ecological *and* an evolutionary side to it. Yale paleontologist Elisabeth Vrba, who has studied these 2.5-million-year-old African events closely, recognizes two different sets of reactions to the global cold snap. On the one hand, we have a purely ecological reaction: The colder, drier climate was no longer hospitable to many of the tropical woodland tree species that dominated the eastern and southern African landscape. Instead the climate was more conducive to grass species, and grasslands therefore spread at the expense of woodlands. Much of this is pure ecological replacement—habitat tracking—as preexisting grass and woody plant species that thrive in drier and cooler climates moved in, and the trees adapted to the warmer, moist climates of the old woodlands died off. This is habitat tracking pure and simple; and many of the mammals that come in to occupy the savannas and grasslands also were simply migrants into the region, having already been adapted to grassland habitats elsewhere.

Second, as Vrba points out, there is an evolutionary response to this ecological disruption as well: Some species almost certainly died out completely. For example, the fossil record yields no remains of our remote ancestor *Australopithecus africanus*, which unequivocally is younger than 2.5 million years, and the inference is fairly clear that that species became extinct as the climate changed. The nearly simultaneous and immediate appearance of two *new* species of hominid underscores perhaps the most important evolutionary aspect of such ecological upheavals: When ecosystems are severely disrupted, habitats become fragmented, often isolating parts of species, and sometimes leading to the rapid evolution of entirely new species.

It has been recognized for over a half century now that new species do not evolve by slow transformation of entire species-lineages through geologic time. Rather they evolve through the isolation of parts of an ancestral species for a sufficiently long period of time to allow changes in the reproductive capacities, and perhaps other adaptations as well, such that the two isolated parts of the ancestral species will no longer be able to interbreed successfully should they ever come in contact again. What was *not* realized until recently is that most speciation events—in a variety of different lineages, all living in the same general area—often occur nearly simultaneously as an evolutionary reaction to momentous environmental change.

Neither speciation nor natural selection-mediated evolutionary change depends entirely on such periods of rapid environmental change and consequent ecological disruption. On the other hand, only recently have biologists come to realize how much evolutionary stability and change do, in fact, hinge on ecosystem stability and disruption, respectively. The physical environment—meaning, for the most part, climate, especially marked and rapid changes in global temperature regimes—seems to drive the entire system.

Until now. Now a single biological species, our species, *Homo sapiens*, is disrupting ecosystems and driving species to extinction in the new wave of mass extinction. Why not just let the Sixth Extinction run its course? After all, evolution ultimately creates new species that become the players in newly rebuilt ecosystems. The answer is simple: New species evolve, and ecosystems are reassembled, only after the cause of disruption and extinction is removed or stabilized. In other words, *Homo sapiens* will have to cease acting as the cause of the Sixth Extinction—whether through our own demise, or, preferably, through determined action, before evolutionary/ecological recovery can begin. Our fate is inextricably linked to the fate of Earth's species and ecosystems.

Chapter 3 THE TREE OF LIFE *How can we comprehend the vast sweep of living diversity, the more than 10 million species now living, not to mention the millions more that have evolved and become extinct over the 3.5-billion-year history of life? How can one's mind possibly grasp all this diversity? Daunting as the task may seem, the natural groups that are the product of the evolutionary process greatly simplify the goal of understanding living diversity.* ❦ *Birdwatchers quickly become intimately familiar with their field guides—and argue passionately about which of the several available is best for identifying the birds of, say, western North America. Though there are over 700 species known from this area, it is still possible to depict and discuss briefly each one of them within the covers of a single guide, which then measures up very closely to the scorecard* you get at the ballpark. No such detail and comprehension can be attempted when we contemplate over 10 million species, with groups that—unlike birds from a single region—are far from being relatively homogeneous, are indeed as disparate as microscopic bacteria and amoebae, slime molds, yeasts, mushrooms, lichens, ferns, sunflowers, sharks, and leopards.

In a more general way, we can chart the basic living groups, along the way noting their fundamental anatomical, physiological, and behavioral features, their numbers and characteristic roles in the ecological arena. The central chart of this chapter provides additional detail, spelling out the major phyla (relatively large-scale natural groupings, or *taxa*) and the latest consensus on their evolutionary relationships (where they fit in life's grand genealogical history). For those seeking more information on particular groups, I have listed in the bibliography a number of crucial, basic references that spell out the biology of these diverse groups in much denser detail.

As Figure 15 (on p. 70) shows, the list of five basic groupings of organisms is itself divided into two parts. As we briefly encountered in chapter 2, the bacteria, lacking true nuclei and other internal organelles ubiquitously found in all other organisms, constitute the Prokaryota. All other organisms are Eukaryota ("whole-celled"), meaning that their main component of DNA is housed in a discrete nucleus and that certain structures, such as mitochondria and chloroplasts (the factories supplying energy to build, maintain, and run each cell's functions) are characteristically present. As we shall see, some bacteria are more closely related to eukaryotes than are others, leading many evolutionary biologists to think of Prokary-

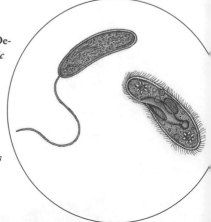

[Figure 14] *A prokaryotic cell (the bacterium* De-scelfovibrio desulfuricans, *left) and a eukaryotic cell (the protoctistan* Paramecium caudatum*). Archaebacteria and Eubacteria are prokaryotic. All the rest of life—the Protoctista, Plantae, Fungi, and Animalia, share eukaryotic cell structure. This is evidence that all are descended from a single common ancestor. This ancestor, in all likelihood, evolved from the fusion of two (or perhaps even more) prokaryotic ancestors.*

ota as more of an informal grouping of similar, simple nucleus-less organisms than as a truly unified group.

Single-celled eukaryotes, sometimes called the Kingdom Protoctista, include familiar names such as amoebae and algae, plus a host of far less familiar groups. These largely microscopic creatures themselves constitute a vast array of disparate forms, and microbiologists, using relatively new techniques of electron microscopy and molecular biology, are only now unraveling the details of their relationships and basic biological features. Nonetheless, enough has already been discovered to make it clear that some of these single-celled eukaryotes are more closely related, respectively, to the multicellular true plants, fungi, and animals. Thus, the evolutionary relationships of protoctists are multifarious, meaning that, like the bacterial prokaryotes, the single-celled eukaryotes are not really a single, evolutionarily homogeneous group. [Figure 14]

True plants, fungi, and animals do appear to be evolutionarily coherent. In other words, all the species within each of these three groups (generally called formally, and respectively, Kingdom Plantae, Kingdom Fungi, and Kingdom Animalia) are descended from a single common ancestor, sharing some common features that mark each of these groups as a coherent evolutionary whole.

These five basic groupings make a fine start to the task of setting up a scorecard of the living world. Let us now examine the contents of each group—starting from the top, with the more complex and familiar organisms, animals and plants, and peeling back layers of evolutionary history until we end with the microbes, which turn out to be the real stars of the game.

ANIMALS To many of us, "animal" is synonymous with "mammal," although, when pressed, we will quickly throw in birds, reptiles, amphibians, and fish. But what is a fly, a shrimp, a garden snail, or, tougher yet, a coral or, hardest of all, a sponge? *All* are animals, meaning multicellular organisms that are *heterotrophic*, procuring their energy from consuming—eating—other forms of life: Other animals to be sure, as well as plants, fungi, and microbes. Most are actively mobile, though some, like corals and sponges, are sedentary. Most of the mobile forms are bilaterally

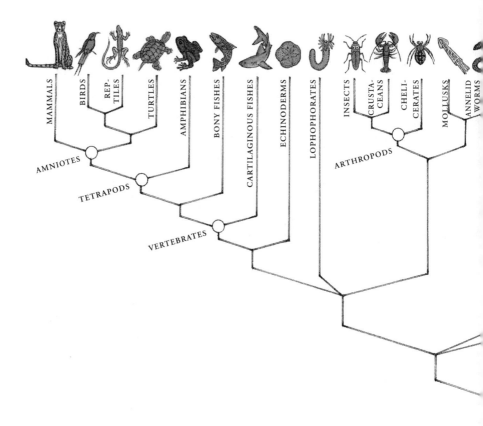

[Figure 15] *A diagram of the evolutionary relationships of all life, organized to match the order of discussion of each of the major groups in this chapter. Though the basic structure of this diagram is unlikely to change, scientists (systematic biologists) are still refining our knowledge of the relationships among living things: This diagram (a cladogram) should be taken as the latest version of current understanding—a theory destined to be tested and refined as knowledge grows. Only recently, for example, has it been suggested that fungi and animals are more closely related to each other than either is to plants—a conclusion that may well be refuted with additional research.*

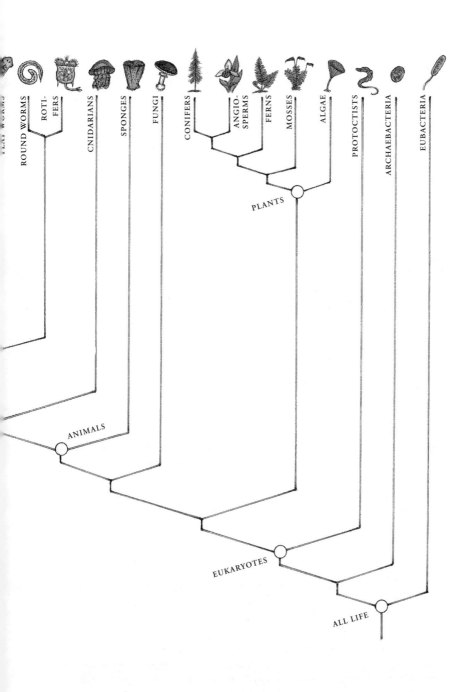

FLAT WORMS
ROUND WORMS
ROTI-FERS
CNIDARIANS
SPONGES
FUNGI
CONIFERS
ANGIO-SPERMS
FERNS
MOSSES
ALGAE
PROTOCTISTS
ARCHAEBACTERIA
EUBACTERIA

PLANTS

ANIMALS

EUKARYOTES

ALL LIFE

symmetrical (their bodies are divided longitudinally into near mirror-image halves) and come equipped with heads with mouths, and bear such sensory apparatuses as eyes and various sound and smell detectors. There are exceptions: Echinoderms such as starfish and sea urchins are built on a fivefold radial plan and lack anything remotely resembling a head.

The main ecological point about animals is that they are always somewhere up along the food chain. Plants and photosynthesizing microbes are always the primary producers, trapping solar energy that becomes the basis of existence for all other living forms. Animals, of course, come in a great variety of shapes and sizes, specialize on an amazing smorgasbord of food types, and thus play myriad roles in ecosystems. There are some 37 different phyla of animals—major divisions of the Kingdom Animalia—with one not recognized and named until the very end of 1995. All but a handful of these 37 phyla are exclusively marine, reflecting the Kingdom's ancient aquatic origins. Only some vertebrates (most mammals, birds, reptiles, and amphibians, and even some fish), arthropods (insects and kin, spiders and scorpions, plus one lineage of crustaceans), certain snails (Phylum Mollusca), earthworms (Phylum Annelida), plus a few less familiar, wormlike creatures (Phylum Nematoda) are fully *terrestrial*, meaning that they live on land and extract oxygen from the atmosphere, rather than from aqueous media.

What follows is a personalized tour of the animal roster—with no pretense to exhaustive completeness, but nonetheless in the spirit of portraying the breadth of diversity in the animal kingdom. I start with our own group, Phylum Chordata, Subphylum Vertebrata—the vertebrates, by far the most familiar to us of all organisms on Earth.

We, species *Homo sapiens*, of course, are vertebrate animals—though of rather a peculiar sort. [Figure 16]

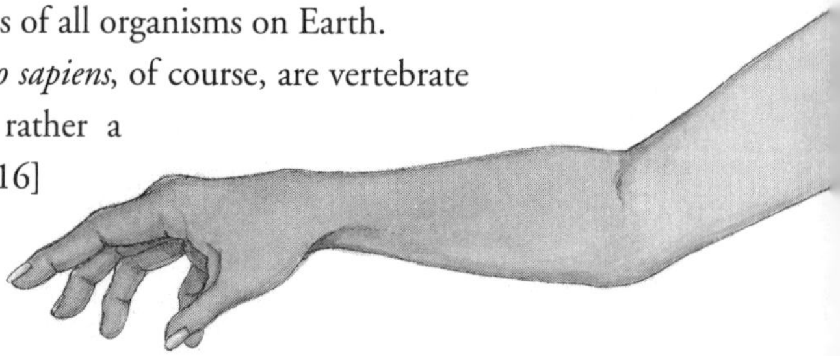

[Figure 16] *Our species,* Homo sapiens, *showing our opposable thumb—one of the unique anatomical specializations that have contributed to our evolutionary success. Despite our huge numbers (6 billion and counting), and our geographically-based variation (so-called "races"), humans are far more genetically homogeneous than the various races of African chimpanzees (the species* Pan troglodytes).

Unlike all other organisms, most humans no longer live within the confines of local ecosystems. Through our self-awareness, our cognition, and our capacity to use language, we have collectively devised an elaborate behavioral system handed down through the generations by learning rather than through genes. This is our unique adaptation through which we perform the usual gamut of animal functions: obtaining food and avoiding predation, disease, and the rigors of the weather.

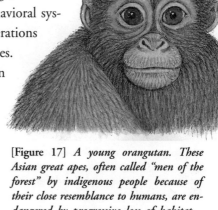

[Figure 17] *A young orangutan. These Asian great apes, often called "men of the forest" by indigenous people because of their close resemblance to humans, are endangered by progressive loss of habitat— primarily through logging of their forest homes, but most recently from the out-of-control fires that have become rampant in much of Malaysia and Indonesia.*

We are members of the Order Primates, allied through a long evolutionary sequence with the great apes, Old and New World monkeys, and the "lower primates," including lemurs and lorises. We are among the more primitive of mammalian groups, stretching all the way back across the Mesozoic-Cenozoic boundary to the Cretaceous, when true mammals first began their evolutionary radiation. [Figure 17]

Other long-lived mammalian groups include the insectivores—moles, shrews, hedgehogs and the like. [Figure 18] Madagascar retains a large variety of species, including the common tenrec, which, at a half meter long, is a veritable giant in shrewdom. Traveling in northeastern Madagascar,

[Figure 18] *The streaked tenrec,* Hemicentetes semispinosus, *an insectivore from Madagascar. Insectivores are among the most primitive of living mammals—living reminders of Mesozoic days when dinosaurs ruled Earth, and insectivores were among the few kinds of mammals present. Though not rodents (which hadn't yet evolved), figuratively speaking the ancient insectivores and other small mammals were the rats of the Mesozoic.*

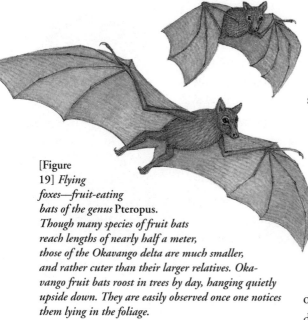

[Figure 19] *Flying foxes—fruit-eating bats of the genus* Pteropus. *Though many species of fruit bats reach lengths of nearly half a meter, those of the Okavango delta are much smaller, and rather cuter than their larger relatives. Okavango fruit bats roost in trees by day, hanging quietly upside down. They are easily observed once one notices them lying in the foliage.*

my companions and I once stopped to buy several squirming specimens from some local kids. *They* were merely trying to supplement their families' meager incomes; *we*, imbued with the knowledge of the rarity and precarious status of this species (and so many others on that ravaged island), duly set the tenrecs free several miles along the road—a microcosm of the conflicting values and uses people have for wildlife. Our gesture, however well-meant, was merely a token and at odds with local perception of rights and needs. Only through concerted effort—especially where conservation of species becomes in the best economic interests of the often impoverished people of a region—can such conflicts have any hope of resolution.

Insectivores, as their name implies, are largely insect (and earthworm) eaters, as are bats (Order Chiroptera), close relatives presumably derived from early insectivores. Bats are one of the most diverse group of mammals, with over 986 species known. [Figure 19] More species are being discovered all the time. A research team from the American Museum of Natural History, including husband and wife Nancy Simmons and Rob Voss, trapped individuals of some 79 bat species—including one that no one had ever seen before—between 1991 and 1994 in a small area of French Guinea. "Flying foxes," which can reach lengths of 40 centimeters and wingspans of 1.7 meters, reflect a switch from ancestral eating habits, from insect to fruit. Anyone who has (as I have done) inadvertently alarmed a tree full of these sleeping giants knows the quick surge of adrenaline, which quickly gives way to pure thrill as these huge bats take to wing in chaotic turmoil.

Rodents, with some 1,814 species known, outnumber bats in sheer diversity of species. Rodents are distinguished by ever-growing incisors, allowing them to gnaw away at their obligate vegetarian foodstuffs. They live in an amazing array of environments: from the arctic tundra (lemmings) to the depths of the tropical rain forest (South America's capybara—the world's largest rodent—and various species of agouti leap to mind). Guy Musser, another colleague of mine at the American Museum, once did a 3-year stint on the Indonesian Island of Sulawesi (the former Celebes), trapping 29 species of rodents, including giant rats. Nearby Sumatra also has its giant rat—made famous in the Sherlock Holmes tale.

Mammalian herbivores belong to several separate large-scale lineages, foremost among which are the odd-toed Order Perissodactyla (horses, rhinos, tapirs, and many extinct groups) and the even-toed Order Artiodactyla (including pigs, sheep, camels, deer, giraffes, antelopes, and cattle). Seemingly, the heyday of the odd-toed group (horses have one and rhinos three on each foot, front and back) has passed. Rhinos used to swarm the Cenozoic American plains, as did the now extinct brontotheres. Stanley Kubrick, director and cowriter of the movie *2001: A Space Odyssey*, used a tapir to symbolize truly ancient times, as he depicted critical events in early human evolution at the movie's beginning. [Figure 20] Never mind that

[Figure 20] *The Asiatic tapir,* Tapirus indicus. *Tapirs today face endangerment, and ultimate extinction, through the rampant destruction of their habitat in both Asia and South America.*

tapirs today occur only in Asia and South America, and, according to the fossil record, never were in Africa with the early hominids. Nonetheless, the ancient *look* of the tapir was just right for the movie.

Most perissodactyls and artiodactyls are grazers, and most live in herds. American bison, tundra-dwelling musk oxen, African wildebeest, and Thomson's gazelles are typical grass-grazing, herd-dwelling species. But many species are browsers of bush, such as the long-necked gerenuk, an East African antelope. In the depths of the West African rain forest, the bongo, a large, camouflaged antelope, lives in small groups and is a browser—as is the equally elusive okapi, a short-necked relative of giraffes not discovered by Western scientists until 1900. Indeed, relatively large-sized artiodactyls are still being discovered in surprising places such as the forests of Vietnam, where several musk deer and antelope species have recently been found and are being described for the first time by my colleagues, biologists George Schaller of the New York Zoological Society and Elisabeth S. Vrba of Yale University.

Elephants—close relatives of the little hyraxes of the Old World [Figure 21]—are still with us, but only as two distinct species, a mere vestige of a once greatly diverse group that roamed over the Earth's continents. Elephants have recently begun once again to collide with humans, as their voracious appetites, which demand large ranges, conflict with ever-increasing human needs for more agricultural land—not to mention the still-unset-

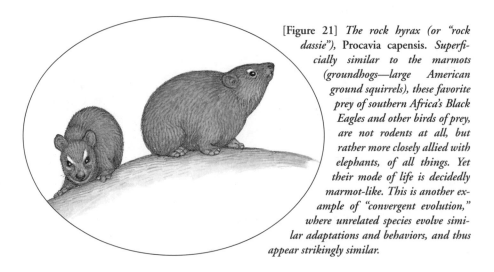

[Figure 21] *The rock hyrax (or "rock dassie"),* Procavia capensis. *Superficially similar to the marmots (groundhogs—large American ground squirrels), these favorite prey of southern Africa's Black Eagles and other birds of prey, are not rodents at all, but rather more closely allied with elephants, of all things. Yet their mode of life is decidedly marmot-like. This is another example of "convergent evolution," where unrelated species evolve similar adaptations and behaviors, and thus appear strikingly similar.*

tled question of the ivory trade. Famed paleontologist and conservationist Richard Leakey recently burned a huge pile of elephant tusks confiscated from poachers over the years; but some think a better approach might have been to flood the market with the ivory, reducing demand and thus lowering prices and making elephant poaching far less attractive. Conservation policy and practice are not always as straightforward a matter as one might wish.

Whales, like bats, mark an abrupt change in direction of mammalian evolution, for both sprang from four-legged terrestrial mammalian ancestors. Whales (and dolphins—Order Cetacea) are fully aquatic, yet retain fundamental mammalian features, such as three middle ear bones, milk production for young, placental development in the uterus, and hair (sparse, but there). Whales simply are not fish, and the fact that they have, in a sense, "gone back" to the sea but retain the major advanced features of their land-living mammal ancestors is dramatic evidence of evolution at work. I have stood at the bow of a ship in the Gulf of California between Baja and the Mexican mainland and watched in awe as a 30-meter blue whale—the largest animal on record—glided across our path. It seemed to take forever. Whales are coming back—thanks to international covenants for their protection and sensible harvesting where total bans have been lifted. Allowing some Inuit peoples the right to take some species (such as right whales), as a traditional source of food and other supplies, is an example of sound integration of conservationist concern with the need to take into account local traditions and economic needs.

Seals, in contrast to whales, are neither wholly aquatic nor members of a mammalian order all their own: They are, instead, members in good standing of the Order Carnivora, which includes as well bears, dogs, raccoons, hyenas, true cats, weasels, and their numerous relatives. High on the mammalian food chain, virtually all of the larger carnivores are under threat. The famous old cartoon that shows a little fish being eaten by a slightly larger one, etc. gives the false impression that for every prey organism, a carnivore is lurking. What really happens is that, the higher up the food chain, the fewer number of individuals are to be found. There are vastly fewer wolves than reindeer and musk oxen on the tundra. Yet it is those wolves, or the Indian tiger, or the lions, leopards, and cheetahs of Africa, the jaguar of South America, the mountain lion of (predominantly) western North America that so rivet our attention. There is absolutely nothing like waking up in the middle of the night in your tent on an African safari, with the entire campsite reeking of leopard scent or the lions roaring 50 meters away, just after they have padded through.

BIRDS Birds are living dinosaurs, meaning that all birds, from ostriches to sparrows, find their closest relatives among the extinct sauropod dinosaurs, which included the likes of *Tyrannosaurus rex* and the huge brontosaurians. But most birds, like bats, fly; lacking teeth, birds pierce, bite, and crush their variegated foodstuffs with their specialized horny beaks. Their feathers, like mammalian hair, were derived from primordial reptilian scales, perhaps first as a means of regulating body temperature (staying warm), and later as indispensable adjuncts for powered flight.

Birds are among the best known creatures on Earth. There are over 9,000 species currently known, and though new species are still occasionally encountered in remote regions, scientific knowledge is perhaps more nearly complete for birds than for any other group of organisms. The main reason is that the vast majority of birds are diurnal—flying, eating, and mating by day. In contrast, mammals to a very great extent remain nocturnal, a hangover from the bygone days of the Mesozoic when virtually all mammals were nighttime skulkers, and dinosaurs, the ancestors of birds, ruled the day.

Though some birds are small or otherwise cryptic and secretive, many are brightly hued and move around in the open. Birds are by far the most conspicuous group of animals, and though bird vision and hearing tend to be far keener than our own, their calls, songs, and often colorful plumage patterns—the very items they use to recognize each other—are both audible and visible to us. Thus, we have a direct window into the bird world—an ability to recognize separate species, separate sexes, and even sometimes individual birds which, ironically, we lack for many mammals, which often rely on a keen sense of smell to recognize one another.

Thus birding, reputed to be the most popular field of amateur science after astronomy, is founded on a natural attraction to the living world, greatly abetted by the relative ease of observation of birds and the fact that their sounds and plumage patterns fall within our own sensory capacities.

Not to suggest that birding carries only scientific pleasures: For many, including myself, birdwatching affords an easy entrèe into the natural world, a way to penetrate unfamiliar ecosystems in faraway places, and a means of simply enjoying the natural world. It is also a sport—hunting without a gun for the sheer enjoyment of it.

Most primitive (meaning closest to the ancestral condition) among all living birds are the large, flightless ostriches and near relatives found only on the continents of the southern hemisphere—suggesting that the group has been fragmented by the breakup of the old supercontinent of Gondwana (Gondwanaland). Though, anatomically speaking, the species of the ostrich group are in many respects archaic, their most obvious characteristic—their inability to fly—is a secondary loss of flight. That an African ostrich is somewhat reminiscent of a two-legged dinosaur from the Mesozoic is evolutionary convergence and a sign that even the most primitive members of a group will invariably have their own evolutionary specializations.

In the northern hemisphere, the loons—marvelous divers, inept clamberers on dry land—are in many ways the most primitive group. They and the grebes fish for a living. Ducks, however, show a much broader range of eating habits; some, like the familiar mallards, are dabblers, eating vegetation in shallow waters (geese, in contrast, graze on dry land). Other ducks, such as the mergansers, with their saw-toothed sculptured bills, dive for fish. The male hooded merganser is my candidate for most beautiful bird of northeastern North America, my home birding haunt.

Raptors—hawks, eagles, falcons, and vultures—symbolize power, majesty, and utterly efficient killing to people around the world. Though their size is often impressive, raptors actually span a large size range, from the huge Andean condor to diminutive falcons. I'll never forget seeing an immature Madagascar fish eagle—a very rare and endangered relative of our own bald eagle—lumbering over a coastal river, absolutely dwarfing the pied crow and an even smaller crested drongo (a black bird about the size of thrush). In sharp contrast, the also rare Taita falcon (in effect, a miniature peregrine falcon), which I once was lucky to see at Victoria Falls in Zimbabwe, is only 27 centimeters long (about the size of a blue jay), and the African pygmy falcon, at 20 centimeters, is smaller than an American robin. The range in body size reflects the way, in an evolutionary sense, a group of carnivores divide up their potential prey, for though there is inevitably overlap, and sometimes very large carnivores catch very small mammals and birds, by and large it is true that large carnivores eat larger prey items, while their smaller, faster, and more maneuverable kin focus on rodents and small birds.

Raptors hardly exhaust the list of flesh-eating birds; owls are in many ways the nighttime equivalents of the hawk-eagle cadre—though some species, like the burrowing owl that haunts prairie dog towns, are readily seen standing outside their burrows in broad daylight, or, as is the case for the widely distributed short-eared owl, flying like a giant moth over open fields at dawn and dusk. Crows and ravens, too, are sometimes considered "honorary" raptors, soaring, as ravens are especially wont to do, like eagles, looking for prey or carrion. Even the perching songbirds make their contribution to the carnivorous predator array: shrikes, often called "butcher birds" for their penchant of impaling small rodents and large insects on thorns for later consumption, are true hunters. Shrikes, like many frog

[Figure 22] *An African taita falcon (*Falco fasciinucha*) and the North American loggerhead shrike (*Lanius ludovicianus*). Though diminutive, the taita is in many ways a typical falcon, a member of a major subgroup of true birds of prey, the Family Accipitridae. Shrikes, in contrast, are the only songbirds to prey on large insects and small rodents. Primarily an Old World group, there are only two species of shrike in the New World.*

species around the world, have been undergoing a mysterious, virtually ubiquitous, decline in recent years—except in Africa, stronghold of a stunning variety of these birds. [Figure 22]

Once we remind ourselves that insects are animals, too, then the ranks of avian carnivores swell dramatically. The flycatchers of the Old World are completely unrelated to those of the western hemisphere: Those of Europe, Asia, and Africa are closely related to thrushes, chats, and Old World warblers, while North and South American flycatchers form a more primitive and utterly distinct group. The parallels in appearance and behavior between the two groups are remarkable, however: Most species of both groups of flycatcher are inconspicuously colored and sit upright and motionless, sallying forth to grab a flying insect and returning to the same or to a nearby perch to repeat the process. Other obligately insectivorous groups are the Old and New World warblers—once again examples of two unrelated groups performing very similar ecological roles in different regions of the world.

[Figure 23] *Three species of Galapagos finch: warbler finch* (Certhidea olvacea, *left), medium tree finch* (Camarhynchus pauper, *upper right) and large ground finch* (Geospiza magnirostris, *lower right). The latter two species are equipped with typical finch-like bills adapted for cracking open seeds of various shapes, sizes and hardnesses. The warbler finch, however, demonstrates how evolution can modify ancestral adaptations—in this case fashioning a bill for insect eating—and thus looking more like a warbler's beak than a finch's beak, more "convergent evolution."*

Most birds are more flexible in their dietary requirements. Woodpeckers specialize on insects, but will eat sunflower and other seeds. The American robin, a species of thrush, is most famous for plucking worms from suburban lawns, but switches to fruits as fall approaches and when winter sees the insect population dropping to nil. The obligate seed eaters, all generally small, are often quite colorful birds with beaks designed for cracking and crushing rather than piercing or biting. Finches of various families, plus the buntings and sparrows, all fill this particular bill. [Figure 23]

Birders can go on forever, and there are many more players in the game of life to consider. But, as the most accessible form of animal life to observe, birds are the form of life with which many people can most readily identify. Birds play more than the usual role of energy transfer within ecosystems (i.e., eating and being eaten) or the spreading of seeds of tropical plant species: Perhaps more than any form of life, they help humans keep in touch with the natural world. Returning to the Galapagos for a moment, imagine seeing (as I have) a quintessentially tropical creature like a flamingo sifting minute crustaceans from a briny pond, while, less than a mile away, penguins, symbols of Antarctica's frozen landscape, cavort along the shoreline. A juxtaposition of opposite worlds, yet all quite natural on an island chain that sits astride the equator, but is bathed by the frigid Humboldt current. Birds can get you thinking about the natural world that surrounds us all.

REPTILES AND AMPHIBIANS Frogs and salamanders are amphibians, smooth-skinned insect-eating descendants of the first vertebrates to crawl out on land, and are thus founding members of the earliest terrestrial ecosystems. Frogs are truly amphibious, with many species spending more time out of the water than in it. What makes an amphibian an amphibian is not so much its daily mode of life, but rather its reproductive habits. All amphibians—frogs and salamanders—must seek out water, or in some instances very moist soils, or even the leaves of very humid tropical plants, to lay their eggs.

Not so reptiles—meaning turtles, crocodilians, tuataras, lizards, and snakes. In all species except for the relatively few that give birth to live young, these organisms possess amniote eggs, in which the developing embryo is ensconced in a protective, water-filled membrane, enveloped in a hard, calcified shell. Bird eggs are very similar to reptilian eggs. Striking additional confirmation that birds are actually members of the reptilian dinosaur clan came recently when a specimen of the small dinosaur *Oviraptor* was found (by American Museum paleontologist Mark Norell and colleagues) in Cretaceous sediments in Mongolia, sitting on a clutch

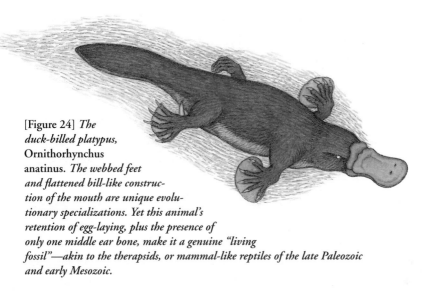

[Figure 24] *The
duck-billed platypus,*
Ornithorhynchus
anatinus. *The webbed feet
and flattened bill-like construction of the mouth are unique evolutionary specializations. Yet this animal's
retention of egg-laying, plus the presence of
only one middle ear bone, make it a genuine "living
fossil"—akin to the therapsids, or mammal-like reptiles of the late Paleozoic
and early Mesozoic.*

of eggs, its legs folded under exactly like a chicken! We mammals, too, are
amniotic: The "water" that "breaks" signaling the imminent arrival of a
new human baby is none other than the amniotic fluids of the ruptured
amniotic sac.

Thus, birds and mammals are both evolutionary spin-offs of a grand
group of scaly, amniotic, egg-laying reptiles. The duck-billed platypus,
which has hair and mammary glands, is a mammal but lays eggs externally
and has only one middle-ear bone, and thus is a vestige of one of the lineages of that branch of early egg-laying, scaly amniotes. [Figure 24] All
modern scale-bearing amniotes—reptiles—are actually more closely related to birds than to mammals.

Save—perhaps—turtles. Turtles are amazing—another great example
of a primitive group that is, nonetheless, highly specialized. Turtles have a
solid skull roof with none of the holes in the side of the head found in the
rest of the amniotes (all other reptiles and birds have two on each side,
while we mammals have but one). That makes them ancient and possibly
descendants of the earliest amniote lineage that led to all other reptiles,
plus birds and mammals.

On the other hand, though I have seen giant ground hornbills (superficially turkeylike, but carnivorous, birds of the African savanna that I have
encountered on the islands of the Okavango Delta) picking out tender

morsels from a tortoise shell, it is also true that Nile crocodiles crunch down to absolutely no avail on the turtles that share their rivers and water holes. The sheer implacability of turtles, their ability to withstand virtually all manner of environmental challenge—from predators to periodic bouts of dryness—can only mean that life in a shell is an evolutionary adaptation that has continued to work for at least 200 million years. Aesop's fable of the tortoise and the hare—with its moral "slow and steady wins the race"—is as applicable to evolution over the eons of geologic time as it is to the more mercurial and ephemeral affairs of humans.

Snakes are another issue. [Figure 25] Though some profess genuine love of snakes (Natassia Kinski did—leading to the famous Richard Avedon photo of her entwined with a python), stereotypically, at least, we humans abhor them—all the while fascinated by them. Rudyard Kipling's description of a band of monkeys enthralled, virtually hypnotized by the great python Kaa, captures that dual sense of fear and fascination that we monkey-relative humans tend to feel about snakes. I think it more than likely that our fear of snakes is primordial and is

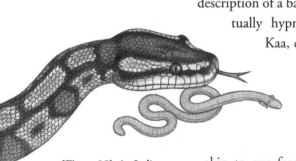

[Figure 25] *An Indian python* (Python molurus), *and the two-legged worm lizard,* Bipes biporus, *an amphisbaenid. The snake-like amphisbaenids are actually lizards—yet another example of convergence in evolution.*

akin to our fear of large cats, larger birds of prey, and fire. Why? Consider our 60-odd million years of living in trees before an ancestral species of ours adopted a bipedal, open-savanna existence, a process begun probably no longer than 5 million years ago. Snakes—along with cats, large birds of prey, and fire—are the principal enemies of arboreal primates. (In Madagascar, I once saw a small band of brown lemurs hurtle to the ground, to our very feet, vastly preferring whatever problems we humans might pose than run the risk of being taken by the large hawk that had just flown overhead.)

Fear of snakes is, I am certain, a stark reminder of the days when we were literally still living "in a state of nature." In a very real sense, what Yale University biologist Stephen Kennert has called the "negativistic" value of such emotions is a clear signal of the deep connection between *Homo sapiens* and the rest of the natural world.

FISH   "Fish" is a loose term; anything in the ocean—including the mammalian whales or even some invertebrates—is a fish to some people. (How many people know that cuttlefish are actually squids—which are mollusks, not vertebrates—and that the cuttlefish "bone" in your bird cage is no bone at all but an internal shell?) Even among the vertebrates, there are several kinds of not particularly closely related "fish": two different sorts of jawless fishes, the hagfish and parasitic lampreys (both remnants of really early stages of vertebrate evolution); the great and fearsome array of sharks, skates, and rays; and a truly amazing array of true "bony" fishes— the only kinds of fish with actual bones in their bodies—herring, carp, salmon, bass, tuna, swordfish, *those* kinds of fish. Real fish.

Sharks and rays have skeletons made of cartilage, which can, indeed, be hardened, impregnated with calcium salts. The result is something like bone, but independently evolved and not true bone at all. Sharks go all the way back to the Devonian, when they swam in seas crowded with the ancestors of our modern fish and with still others that have since gone the way of extinction. As nearly anyone who has gone to the movies or watched the fearsome TV videos knows, sharks are master hunters; some species are able to detect incredibly small concentrations of blood over long distances. Though we find them fascinating, and though some species, such as mako, even make it to the dining table, sharks no longer loom as the most important vertebrates of the world's oceans.

That title belongs to the true bony fish. There are some 25,000 species of bony fishes known, and more are being discovered all the time. Though freshwaters around the world abound in species, most bony fishes are marine, and most of these are tropical. The array includes some marvelous remnants of old branches of fishdom—one of which led to us terrestrial vertebrates (known in the trade as "tetrapods" or "four-legged ones"). These are lungfish (which, like the ostrich group, today has a Gondwana distribution, although fossil lungfish are known from around the world). Some biologists regard lungfish—with their notable capacity to survive periods of drought sealed up in burrows in the desiccated bottom muds of lakes and rivers—as the sister group of the tetrapods. The more conventional candidate for our closest living fishy relative is, however, *Latimeria chalumnae*, the famed coelacanth—like lungfish, a veritable living fossil long thought to be extinct until one surfaced in a fisherman's net along the South African coast in the 1930s. These fishes are lobe-finned, with the bones inside their pectoral and pelvic fins (read "arm" and "leg," respectively) bearing a close correspondence to the arrangement of our own limbs, a sure sign of close evolutionary kinship. [Figure 26]

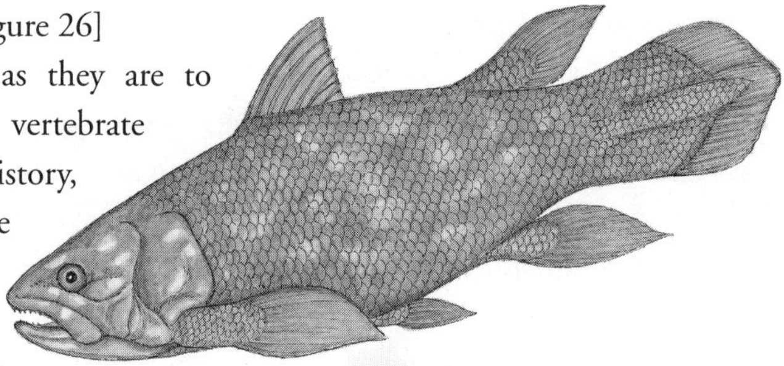

Important as they are to understanding vertebrate evolutionary history, these groups are nothing compared to the typical bony fishes, the so-called actinopterygians. Even here we have some primitive remnants of early days—fish like gar-pike and sturgeon. The evolutionary explosion of the advanced actinops, the *teleosts*, in the Cretaceous Period around 100 million years ago produced the initial spurt of what is now a prodigious array of true bony fish. These range, as we have just seen, from anchovies and herrings up through giant tunas and include the sorts of colorful fish you see in aquaria—most of which come from rivers in tropical climes— as well as the riot of colorful diversity of fishes found in the vicinity of most of the world's coral reefs, which are themselves restricted to the marine Tropics.

[Figure 26] *The coelacanth,* **Latimeria chalumnae.** *Coelacanths first appeared in the Devonian Period of the mid-Paleozoic. Like their equally-primitive kin, the lungfishes, coelacanth diversity was greatest early in their history.* **Latimeria chalumnae,** *the last living remnant of a once important group, is on the brink of extinction, threatened by over-fishing brought on in the excitement of its relatively recent discovery.*

But it is the fishes of the great fishing grounds that concern me most at the moment. The great fisheries lie for the most part in zones of upwelling, where phosphorous and other nutrients are brought from the depths to the surface, supporting a dense population of bacteria, microbes, and tiny animals we call *plankton*, the energy source supporting (in big fish eats small fish eats still smaller fish fashion) a dense array of larger invertebrates and vertebrates—meaning fish and their mammalian predators.

Commercial fishing is the last vestige of the hunter-gathering mode of life that humans otherwise have forsaken since adopting an agricultural mode of existence. And we are very good at it. The Aleut people living on the Pribilof Islands have begun to notice the impact that international commercial fishing is having on their *terrestrial* ecosystem hundreds of feet above the Bering Sea. (The Bering Sea lies north of the Aleutian chain of islands, a volcanic arc that extends toward Siberia from the Alaskan coast. The Pribilofs consist of two inhabited islands, St. George and the larger St. Paul, which are isolated outposts some 400 kilometers west of the Alaskan mainland.)

The situation in the Pribilofs is a microcosm of the plight of marine fisheries virtually everywhere. The Bering Sea is one of the world's most productive and species-rich marine habitats. The fisheries are historically rich, and support, in addition to indigenous human life, large populations of whales, fur seals, other mammals, and seabirds. Indeed, the Pribilofs are the sole breeding ground of the northern fur seal, which, during the winter, travels south along the Pacific coast of North America.

The Aleut people are originally from the Aleutian chain, only populating the Pribilofs for the last 200 years as harvesters of fur seals, first for Russian overlords and later (after Secretary of the Interior William Seward engineered "Seward's Folly," the purchase of Alaska in 1867) under continued harsh supervision by the United States government. Until the end of World War II, the Pribilof Aleuts lived in virtual serfdom, subsisting on meager incomes and very tight rations. They were able to supplement their rations by fishing, hunting, and using the eggs of seabirds, which breed by the thousands on the extensive cliff faces of the two islands.

In 1983, in the culmination of an international outcry against the harvesting of seal pups, the United States government imposed a ban on all but subsistence seal harvesting, a move that forced the Pribilof Aleuts to reinvent their economy. They did so very successfully, by entering the service sector of the international fishing effort in the Bering Sea. The village of St. Paul today has two fish processing plants: one on shore, and the other on a permanently anchored ship tied to a wharf in the harbor. In the winter, dozens of fish processing ships anchor in the lee of St. Paul's northeastern peninsula, seeking some relief from the stormy seas of the perpetually dark Arctic winter. Today, the village of St. Paul has one of the highest per capita incomes of any town in Alaska.

But all is not rosy. The seal populations are down, despite the moratorium on harvesting (they are still being taken by non-U.S. interests, especially as they travel through the Aleutian chain on their annual migrations). In addition, the success ratio of breeding has declined in many of the species of seabirds nesting on the islands' cliff faces. For example, St. Paul and St. George have the only known nesting sites of red legged kittiwakes (kittiwakes are open ocean gulls). The United States government maintains a large complement of biologists on St. Paul whose job it is to monitor the status of all the wildlife populations. One of their more sobering recent findings is that they could find no evidence of even a single successful hatching of a red-legged kittiwake chick in 1994 on St. Paul.

What's going on? Why have these land-breeding species declined? The seals and kittiwakes are serving as a kind of "miner's canary" early-warning system, sending a message, not just about the terrestrial ecosystem on St. Paul, but especially about the devastating effect that the fishing industry is beginning to have on the Bering Sea marine ecosystem itself. The seals and kittiwakes eat fish, and the fisheries have become shockingly depleted.

Pollock is the main species of fish targeted by the Bering Sea fishing fleet. With what seems at first to be a triumph of efficiency, huge mechanized trawlers, dragging gigantic nets over the sea bottom, in effect perform the marine analogue of the forestry practice of clear-cutting a woodland: Literally everything alive is netted, leaving a barren swathe in its wake. Worse yet, all but pollock and a few other commercially viable species are tossed back—dead, of course—into the sea.

The Bering Sea is being overfished, and the ecosystem itself is in danger of collapse. This is clearly not a good situation: It is not good for the ecosystem itself, which means that it is not good for the local people who depend on it both for direct subsistence and economic well-being. Nor is it good for the fishing industry and the hundreds of millions of people accustomed to having fish in their diet.

Nor is the problem restricted to the Bering Sea. The mid-1990s has seen a moratorium on fishing on the Georges Banks off the coast of Newfoundland, posing great economic problems for the involuntarily idled fishermen. The notion of sustainable use—of harvesting renewable sources wisely to avoid depletion—is all the more critically important now, as demand for these resources increases (driven by an addition of 90 million people a year to the world's population) and as technology produces ever more efficient mechanisms to harvest those resources.

INVERTEBRATES

Invertebrates are all those animals that are not vertebrates, which is to say that invertebrates are all of the 36 animal phyla other than the Subphylum Vertebrata of the Phylum Chordata. The distinction is really one between the "haves"—the vertebrates with their backbones—and the "have-nots," a profuse array of animal life ranging from the very primitive sponges through the complex mollusks, arthropods, and echinoderms. Vertebrates are united, in an evolutionary sense, by the simple fact that all have such a backbone. Needless to say, *not* having a backbone conveys no sense of unity at all, and we must look at the details of the anatomies of all these disparate invertebrate groups to detect who is most closely related to whom.

It is not so much adult anatomy, but rather details of the early development of the fertilized egg, which help us to find which among these varied invertebrates are our closest relatives. Within our own Phylum Chordata, we have two groups: the obscure and wormlike hemichordates and the much more familiar tunicates (salps or "sea squirts"), which, with their sacklike bodies attached to seaweed or dock pilings, may ring a bell with beachcombers but hardly suggest any close kinship with ourselves. Yet they do have a nerve cord running along the top of the body, and other features sufficient to unite them within our own Phylum Chordata.

Harder yet to swallow is that the major animal phylum most closely related to our own is the Phylum Echinodermata—to my mind the oddest group of complex animals on Earth. Indeed, they have often struck me as downright other-worldly: True animals, echinoderms (meaning "spiny skin") completely lack the remotest semblance of a head. Most have body plans laid out in a fivefold, radial fashion. Think, for example, of a starfish: The quintessential echinoderm with five (or multiples of five) "arms," a mouth located on the bottom side where the legs converge, no head at all,

[Figure 27] *Sea otter (*Enhydra lutris*) with sea urchin. Sea otters are found along the entire Pacific Rim, from the Japanese Island of Hokkaido to Baja California.*

and no eyes, antennae, or any other conspicuous form of sensory organs, a starfish looks nothing like what we all imagine animals *should* look like. Yet they truly are animals, sharing—with all other echinoderms, plus ourselves and several lesser, wormlike groups—fundamental patterns of early embryological development.

Starfish are eminently successful predators—often the bane of scallop fishermen—as they seize clams with their suckerlike tube feet, wrenching the shell open just enough to extrude their stomachs through their mouths, into the clam shell to begin the process of digesting the clam before it is even dead! The crown-of-thorns starfish, a predator on corals, underwent a population explosion on the coral reefs of the South Pacific in the 1970s, worrying biologists that the reefs would decay if too many of the corals were killed at one time. Like many another population explosion, this one subsided as quickly as it had arisen, and the corals immediately began their natural rebound. Is the current human population explosion likely to mirror the crown-of-thorns event?

Sea urchins—globular, spiny echinoderms—are an important component of an important drama on the Pacific Northwest coast of North America, a story that has important ecological and conservation overtones and involves a keystone species. *Keystone species* are organisms that play such a critical role in a local ecosystem that, once they are removed (for whatever reason), the entire system changes. In this case, the keystone species is not an echinoderm, but rather sea otters, large, playful, and attractive mammals that live around extensive kelp beds (kelp are a species of brown alga).

Sea otters are easy to approach, and their thick furs have been a target of hunters for centuries—so much so that, until very recently, they have been hovering on the brink of extinction. [Figure 27] Now here is how sea urchins enter into the fray. One particular species of sea urchin dines exclusively on kelp. Sea otters, in turn, dine on a rich assortment of Pacific invertebrates, including abalones (which are primitive snails), clams, and—you guessed it—sea urchins. When hunting depressed sea otter communities—totally eradicating them over much of their primordial kelp-bed range—the sea urchin population, no longer kept under control by sea otter predation, exploded. As a result, the kelp beds declined because *their* predator was suddenly present in much larger numbers.

Now, healthy kelp beds are prodigious affairs. They are really nearshore marine forests, with stems as long as 10 meters rooting each algal individual to the sea bottom. Kelp stands are densely packed, and, like true forests, are home to a wide variety of species that hide, hunt, and reproduce—in short, *live*—there. Kelp beds are mini-ecosystems, essential elements of productivity for Pacific Northwest marine ecosystems in general. No sea otters means too many sea urchins, which spells trouble for kelp, and so these mini-ecosystems crash. The sea otters get the accolade of keystone species because their demise led to the crash of the kelp bed communities; but the sea urchins play every bit as critical a role in the story as the sea otters. For my money, the real story is the kelp itself, without which these mini-ecosystems would have no physical basis. In any case, the story is timely, as seldom is the critical relationship between a series of species in an ecosystem so clearly seen. It is good to be able to report that, thanks to an aggressive protection and restocking program, sea otter populations are bouncing back—bad news for the sea urchins, but decidedly good news for the kelp bed communities—and thus for the entire marine ecosystem of the Pacific Northwest.

THE PROTOSTOMES We and the echinoderms are part of a large evolutionary lineage known as the *deuterostomes*. Our mouths are formed, not from the first, but the second, opening that appears in the very early, hollow-ball-like stage of embryological development known as

the *blastula.* In contrast, in *protostomes,* that first opening becomes the mouth. There are other differences as well, but the main point is that we deuterostomes, all in all, constitute the lesser of two unequal branches of complex, multicellular animal life: The protostomes are far more prodigious in overall diversity. The protostome branch of complex animals includes (along with the molluscan and annelid worm phyla) the arthropods—Phylum Arthropoda—easily the most important group of advanced animal life from the point of view of sheer numbers, diversity, and ecological impact. Their name means "jointed legs," referring to their generally firm outer skeleton, and their many legs (plus, in most cases, antennae) divided into several segments. Insects are arthropods, as are crabs and shrimp, spiders, scorpions, and many less familiar groups. Arthropods have invaded virtually every known habitat on Earth. [Figure 28] Although no one knows for sure how many arthropod species there are, nearly 1.1 million species are known, of which 950,000 are insects, and 350,000 of the insects are beetles!

Just how important are arthropods to the natural economy? Harvard University ant specialist and biodiversity expert E.O. Wilson, in his book *The Diversity of Life* (1992, p. 133), has written:

So important are insects and other land-dwelling arthropods that if all were to disappear, *humanity probably could not last more than a few months* [my italics]. Most of the amphibians, reptiles, birds, and mammals would crash to extinction about the same time. Next would go the bulk of the flowering plants and with them the physical structure of most forests and other terrestrial habitats of the world. The land surface would liter-

[Figure 28] *Rhinoceros beetle*—Oryctes rhinoceros. *Probably the longest standing joke about biodiversity is a comment generally attributed to British biologist J.B.S. Haldane (but quite possibly apocryphal). In answer to the question "What has the study of biology taught him of the nature of the Creator?," Haldane is supposed to have quipped "An inordinate fondness for beetles"—sly reference to the indubitable fact that insects are by far the most diverse group of animals, and beetles are by far the most diverse group of insects.*

ally rot. As dead vegetation piled up and dried out, closing the channels of the nutrient cycles, other complex forms of vegetation would die off, and with them all but a few remnants of the land vertebrates. The free-living fungi, after enjoying a population explosion of stupendous proportions, would decline precipitously, and most species would perish. The land would return to approximately its condition in early Paleozoic times, covered by mats of recumbent wind-pollinated vegetation, sprinkled with clumps of small trees and bushes here and there, largely devoid of animal life.

A stark scenario indeed! But one with the ring of truth, as insects in general are so fundamental to the normal processes of pollination, consumption (recycling) and even of true decay. In particular, termites (as we have seen in some detail in chapter 1) are hosts to cellulose-digesting fungi and microbes in their hindguts, providing a major pathway of decomposition of dead, woody tissue, and are the *only* such pathway in the arid Tropics.

Insects are the very symbol of biodiversity. I have peered over the rail of a ship threading its way along narrow channels around the Island of Marajo, where the Amazon makes its final run to the South Atlantic, and watched entranced as gorgeous iridescent blue morpho butterflies sent glimmers of light as they flitted along the thickets at the water's edge. [Figure 29] Morphos are endangered, victims of overzealous collectors; their beauty and increasing rarity render them as icons of the current biodiversity crisis.

[Figure 29] *Two more insects: the blue morpho butterfly* (Morpho didius) *and a lightning bug* (Photuris pennsylvanicus). *The diversity of insects is so vast that it will take armies of systematic biologists, working well into the future, to find and study them all. Meanwhile, as habitats—especially in the tropics—fall to fire and chain saw, thousands of insect species are being lost each year, doomed to extinction before anyone can even mark their existence.*

Not just butterflies, but crickets, cicadas, and fireflies on summer nights, bees buzzing over a field of wildflowers, or, on a nastier note, swarms of blackflies making life miserable on an excursion to the north woods in springtime are merely the more noticeable manifestations of an incomprehensibly diverse array of insect life. Insects burrow, walk, fly, and, in some cases, swim; insects suck plant juices, eat leaves, digest wood, drink vertebrate blood, or eat each other. Insects pollinate plants, and some destroy plant life, including, of course, the domesticated plants that form the core of humanity's agricultural enterprise. For the most part passing unnoticed, insects are by far the most consequential of the higher animals, without whom, as E. O. Wilson has said, life as we know it would quickly come to a halt.

Everyone has their favorite, and mine among the arthropods are horseshoe crabs. Why? Because they are, in many ways, so reminiscent of, so similar and closely related to my long-extinct trilobites, gone these past 245 million years. [Figure 30] Horseshoe crabs are superb examples of ecological generalists—able to survive in a wide range of temperatures and salt concentrations, and dining on a large array of shellfish and worms. They are true living fossils, meaning that their external body form has changed little for over 200 million years. Theirs is the kind of success measured in sheer survival rather than, say, the success of beetles, which is measured in hundreds of thousands of different species.

Horseshoe crabs are not crabs at all; indeed, they are not even crustaceans (shrimp, lobsters, barnacles, true crabs, etc.). In "*Limulus*, an

[Figure 30] *Two very different types of crab: the horseshoe crab (*Limulus polyphemus*), and the ghost crab (*Ocypode quadrata*). Though members of different lineages of arthropod evolution since the early days of the Cambrian Period over 500 million years ago, both horseshoe crabs and true crabs share a crucial piece of behavior: All of them must shed their hard outer skeletons in order to grow, a process that temporarily leaves them vulnerable to predators.*

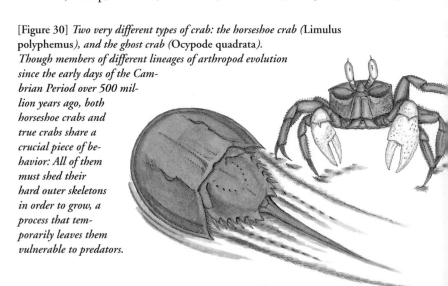

arachnid," a short, classic note in the journal *Nature* in 1885, British biologist E. Ray Lankester summoned all the evidence pointing to affinities between horseshoe crabs (sole remaining members of a much larger group of marine arthropods) and spiders and scorpions. Chief among the features Lankester cited is the presence of *chelicerae*, a distinctive pair of jointed appendages lying in front of the mouth and almost always bearing pincers. True crustaceans always have antennae in the same position. The very essence of evolutionary detective work, reconstructing genealogical relationships based on the shared possession of distinctive traits in an approach generally known as *cladistics*, is beautifully expressed in Lankester's pithy little paper on *Limulus*, a paper written decades before the principles of cladistic investigation had been formally spelled out.

[Figure 31] *The red-gilled nudibranch,* Coryphella rufibranchialis. *One of several groups of snails to lose their shells during their evolutionary history (terrestrial slugs are another), nudibranchs are some of the most beautiful and graceful of all sea creatures. Ironically, though the molluscan shell seems in many ways the key to their evolutionary success, reduction and even loss of that external shell is a common pattern in molluscan evolution, as exemplified especially by octopi and squids.*

Mollusks are another richly diverse protostome phylum, which, if not as central as arthropods in controlling the world's natural economy, nonetheless plays a conspicuous, rich, and varied role in the world's ecosystems. Mollusks are captivating—sort of the invertebrate equivalent of mammals and birds. Most are marine, though some have invaded the terrestrial realm (land snails and their slug cousins; escargot are land snails) and freshwaters (snails and freshwater clams—the latter the original source of genuine "mother-of-pearl" buttons before plastic was invented). What a varied lot they are: lithe, often highly colorful sea slugs [Figure 31]; shelless snails that swim very like manta rays; the more familiar shelled snails, clams, oysters, scallops, tusk shells, chitons; and, perhaps the most astonishing group of all, the cephalopods ("head foots"), the octopuses and squids, some of which represent pinnacles of evolution every bit as complex and fantastic as any mammal or bird species.

Amateur *malacology* (shell collecting) is a major avocational interest in biological natural history, ranking second only to birding. Snail shells are the predominant focus. Cultures in the western Pacific have long used cowrie shells as currency, and the rarity and beauty of cowries, cone shells, tritons, whelks, strombids, and many other snail groups has driven their value on the open commercial market into the many thousands of dollars—a good situation for collectors fortunate enough to own them, or with the monetary wherewithal to acquire them, but problematic for hapless museum curators trying to build and maintain representative collections of as many molluscan species as possible. Fortunately, many shell collectors are aware of the scientific importance of their specimens and are often willing to share or donate scientifically valuable specimens, including some belonging to species not yet described and named in the scientific literature.

New species of mollusks and other invertebrates are turning up all the time. I even discovered one myself, as a student in 1968 at Bodega Bay marine laboratory, on the California coast some 96 kilometers north of San Francisco. There, in a tide pool exposed at low tide at 5 a.m., were two gorgeous, 2-centimeter-long white nudibranch snails with red rhinaria (circlets of gills), belonging to a species we could not identify in all the guide books, monographs, and professional papers at our disposal. If new species are found in the intertidal zone of a populous region like northern California, think of what undiscovered creatures lurk in the nooks and crannies beneath the waves, or on the floor of the abyss 3 kilometers down—and in the newly discovered deep-sea vents. Indeed, the recent discovery of a new species, *Symbion pandora*, living on the mouth parts of North Atlantic lobsters (*Homarus americanus*), made zoologists realize that they had encountered an entirely new phylum of animals dubbed the Cyclociliophora and related more closely to bryozoans (see below) than to mollusks. The point is obvious: The oceans are a treasure trove of as yet undiscovered life. We are a very long way from knowing all there is to know about living biodiversity, about the roster of players in the game of life.

Of all the molluscan groups, though, I find the cephalopods most captivating: Their astonishing vertebratelike eyes (a marvelous example of convergent evolution), streamlined bodies, and jet-propelled swimming make squid, octopuses, and their close kin the most dynamic of invertebrate predators. The jet propulsion is real: A squid traps a volume of water with its mantle flap, expelling it with great force through its *hyponome* (a tube that can aim), and off shoots the squid in the opposite direction, sometimes leaving behind a cloud of ink if it is fleeing a predator.

Cephalopods are pretty smart, too. At Bodega Bay Marine Laboratory, I once saw a tiny octopus skulk around the backside of a small cluster of rocks to come up behind a couple of unsuspecting shrimp of the same size. Pointed in the opposite direction, the shrimp never saw the menace until it was too late for one of them.

My fondest run-in with the world of cephalopods came as a 19-year-old kid, beachcombing near a town in northeastern Brazil, where I was on an anthropological training field trip. I found some small white, loosely coiled shells that I soon recognized as belonging to some kind of cephalopod, as they bore a striking resemblance to the shell of the living species of *Nautilus* [Figure 32], sole living remnant of a once vastly more varied array of externally shelled cephalopod: Nautiloids and their close kin, the ammonoids, the latter now completely gone and only known through their fossilized remains.

All these ancient shelled cephalopods share a fundamental characteristic with their living-fossil relative, *Nautilus*, several species of which survive in the tropical waters of the western Pacific. All have the shell divided into mathematically precisely spaced series of partitions, connected by a tube that regulates the gas and water content—and hence the buoyancy—of the shell. On the beach in Brazil, I had found coiled shells with just those par-

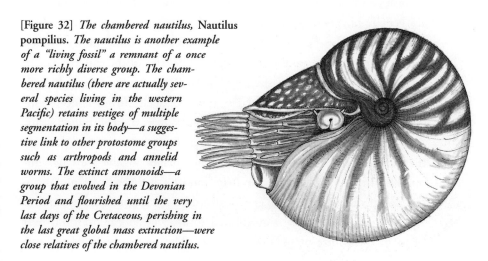

[**Figure 32**] *The chambered nautilus,* **Nautilus pompilius.** *The nautilus is another example of a "living fossil" a remnant of a once more richly diverse group. The chambered nautilus (there are actually several species living in the western Pacific) retains vestiges of multiple segmentation in its body—a suggestive link to other protostome groups such as arthropods and annelid worms. The extinct ammonoids—a group that evolved in the Devonian Period and flourished until the very last days of the Cretaceous, perishing in the last great global mass extinction—were close relatives of the chambered nautilus.*

titions, replete even with a tube system running throughout, connecting the tiniest chamber at the earliest coil all the way through to the end.

What had I found? At that point, all I had had was an introduction to historical geology as a sophomore at Columbia College, where we had spent several lab periods (to my utmost thrill) examining real fossils, including some ammonites and nautiloids. I had to find out what my beachcomber prizes were—a quest that supplied part of the impetus to take a full course in invertebrate paleontology the following fall when I returned from Brazil.

Well, it turned out I was right: My little white shells did belong to a cephalopod species. To my (continued) astonishment, they belonged, not to an externally shelled species, like the nautiloids and ammonoids they so closely resembled, but rather to a species of squid belonging to the genus *Spirula*. Amazingly, the shell in *Spirula* is embedded within the tough fleshy body of a squid, just the opposite of the shells of nautiloids and ammonoids which are housings for the soft body organs. (In direct reference to the magnificent home the nautilus shell provides its owner, Oliver Wendell Holmes wrote in *The Chambered Nautilus*, "Build thee more stately mansions, O my soul.")

Why does the shell of *Spirula*, a squid, so closely resemble the external shells of ammonoids and nautiloids? Though it is possible that the Spirula shell is an actual vestige, an evolutionary remnant, of the shell of an externally shelled ancestor from the Paleozoic, the more likely answer seems to lie, at least in part, in the strong association between molluscan evolutionary genetic history and the mathematical abstraction known as the *logarithmic spiral*. The shells of all mollusks (even clams, which have two such shells) are essentially spirally coiled tubes. The edge of the coil expands in such a way that the overall shape of the shell remains the same, through coil after coil. The penchant for this mode of shell growth is deeply ingrained genetically in mollusks and goes back at least a half billion years. Such an evolutionary condition in effect sets limits on variation. Although there are literally thousands of different shell shapes possible with logarithmically spiralled growth—as the living and fossil diversity of molluscan shells amply illustrates—such genetically imbued mathematical constraints virtually ensure that very similar shell forms will appear over and over again in the course of molluscan history. The more closely related two

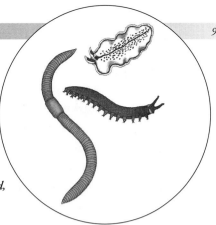

[Figure 33] *Three very different kinds of "worm."*
*At left: an earthworm (*Lumbricus terrestris, *an*
annelid); upper right: a marine flatworm
*(*Psendoceros montereyensis, *a member of the*
Phylum Platyhelminthes; see p. 99); lower right:
*a velvet worm (*Peripatus *sp. member of the Phy-*
lum Onychophora, p. 99). The word "worm" has
been applied to many completely unrelated
groups—in fact, to any of the many soft-bodied,
elongated kinds of animals that are not snakes.

groups may be—such as the squids, on the one hand, and the externally
shelled cephalopods on the other—the more likely that very similar shells
will evolve more than once.

Mollusks and arthropods alone do not exhaust the riches of the vast
protostome branch of complex invertebrate animal life. Among others,
there are still the segmented worms—Phylum Annelida—to consider. Marine members of the annelid phylum share with the otherwise very dissim-
ilar mollusks a nearly identical larval form, the barrel-shaped *trochophore*
larva, which is unmistakable evidence that, despite the vastly different ap-
pearances of adults in both groups, they are nevertheless very close evolutionary relatives. [Figure 33]

Anyone who has seen Humphrey Bogart covered with leeches as he
pulled his stranded vessel through the shallows—with Katherine Hepburn
poling away—in the movie, *The African Queen,* has seen an annelid
worm—as has anyone who has gardened, scoured a leaf pile for fishing
bait, or just walked around after a summer rain, for earthworms are also
annelid worms.

Indeed, earthworms are a powerful force in terrestrial ecosystems, so
important that, late in his life, Darwin devoted an entire book to them. Soil
is a combination of small flakes of rock-derived minerals and organic materials in various states of decay. Earthworms are responsible for renewing the
earthiness of the soil, adding nutrients as dirt passes through their guts, and
turning over the topsoil as well as aerating it as they emerge at night to leave
their castings on the surface. Along with simpler, wormlike nematodes (yet
another major phylum of invertebrates), earthworms are largely responsible
for the rejuvenation of soils, the very underpinnings of terrestrial ecosystems of all types, from grasslands to forests. Although modern high-tech

agriculture has substituted fertilizers and harrows for the traditional, natural dependence of the action of earthworms, nonetheless these creatures remain vital to global agricultural production.

Polychaetes—marine worms—also belong to the great annelid cadre. "Errant" polychaetes scramble across the sea bottom, while sedentary, tube-dwelling species just sit there, periodically extruding a bright array of tentacles that they use to absorb oxygen and filter out nutrients and small bits of edible organic material from the surrounding seawater. The mobile polychaetes have short setae projecting from their side, which they use to gain purchase as they wriggle along the sandy substrate.

It is rare to find species that seem to be evolutionary intermediates—connecting links—between different phyla. The aforementioned tro-chophore larvae, shared by annelids and mollusks, are ample testimony to their affinities, and details of even earlier stages of embryological develop-ment, as we have seen, unite these two groups with arthropods and several other lesser phyla into one major lineage, the protostomes. Adult forms of different phyla are rarely all that similar to one another. A major exception to this generalization lies in a small group—usually given full phylum sta-tus in its own right—the velvet worms, or Phylum Onychophora.

Like the ostrich group among birds, onychophorans have a basically Gondwanaland distribution, meaning they are restricted to continents of the southern hemisphere, where they are usually found in forest settings, living under rotting logs. Velvet worms, as their name suggests, have elon-gated bodies, with an outer body look suggesting segments (like true an-nelid worms), but with an inner body *not* partitioned, but rather opened up as a cavity—like true arthropods. Their lobelike legs, too, seem to be a bridge between the setal "legs" of annelids and the true jointed appendages of arthropods. Like the famed *Archaeopteryx*—the 140-million-year-old fossil, which, with feathers, is an early bird, but with a long tail, and teeth in its head, is distinctly reptilian—velvet worms seem to be mosaics of an-nelid worm and arthropod characteristics.

PROTOSTOME-DEUTEROSTOME ROOTS Disparate as they are, protostomes and our own megalin-eage, the deuterostomes, converge in the dim reaches of evolutionary time,

or in the modern world in survivors of the early days of invertebrate evolution. Protostome and deuterostome animals share a fundamental feature, a
so-called coelom, an internal body cavity lined with cells derived from the
mesoderm, the middle of three layers of cells in the developing embryo.
Though the coelom develops differently in the two groups, nonetheless
prevailing opinion has it that the coelomate animals are all descended from
a common ancestral stock.

What that stock might have been like is suggested by the lophophorates, a group of four or five phyla that seem to display a mixture of protostome and deuterostome traits (though the weight of opinion seems to be
shifting these days toward seeing lophophorates as primitive protostomes).

Brachiopods and bryozoans are the dominant lophophorates. Brachiopods were once the most common form of marine "shellfish" life:
With two shells, they look very much like molluscan clams but are far
more closely related to the bryozoans. The key feature linking these
groups, and separating them from mollusks and other phyla, is the
lophophore, a mass of tentacles extruded, somewhat in the fashion of filter-
feeding worms, to secure nourishment and oxygen from the surrounding
seawater. Because the newly discovered Cyclociliophora (which live on
the lips of lobsters) also possess a circular array of tentacles that look especially like the circular lophophores of the two phyla of bryozoans ("moss
animals," small colonial creatures most commonly seen as filmy encrusta-
tions on seaweeds and dock pilings), the early word is that the Cyclociliophorans are probably another phylum belonging to the lophophorate
group.

PSEUDOCOELOMATES AND BEYOND: TOWARD THE BEGINNINGS OF KING-DOM ANIMALIA

Flatworms (Phylum Trematoda), to my
mind, are the wildest creatures of the animal kingdom. They are some of
the niftiest, even cutest, animals that I have ever seen; they are also among
the nastiest parasites that have ever lived. That's what I mean by "wild"—
these animals inspire a huge emotional swing from admiration, even affec-
tion, to unmitigated disgust, depending on exactly *which* trematode you
are looking at.

First, the good news. *Dugesia* is a genus of free-living flatworm commonly found along rocky shorelines all over the world. With a broad, arrow-shaped head, two eyes above, and a mouth below, followed by a thinner and very flat body, there is no problem seeing *Dugesia* as an animal in good standing. These little guys prowl the intertidal zone, moving around in search of bits of food. Animal psychologists have run experiments testing the intelligence of *Dugesia* and related flatworms and have found that they can learn to navigate a maze, if not as quickly as a rat, then at least with more than a smattering of learning capability—one form of intelligence. I have spent hours just watching these little creatures navigate and have been totally seduced by their *je ne sais quoi* charm.

So much for the good news. Trematodes also include wormlike forms that are, unfortunately, more familiar to humans than their cute little marine relatives. I refer to tapeworms. Tapeworms hook themselves into the guts of humans and other species (each generally specializes on a particular host species) and can grow in humans to lengths in excess of 10 meters. Tapeworm infection is pandemic in the human population—especially rife, of course, in poverty-stricken Third World nations, where sanitation and medical services are far below the standards enjoyed by the industrialized nations. (Those of us who are more fortunate tend to take the availability of nutritious food, safe water, sanitary waste disposal, and medical services very much for granted, while they are increasingly rare luxuries available to the decreasing minority able to afford them.)

Trematodes are the most conspicuous of an eclectic group of animals that occupy a gray zone. Lacking the true body cavities (*coeloms*) of the higher animals (though some, like the trematodes, have body cavities called *pseudocoeloms*, meaning that their development is not the same as in the high, true coelomates), they nonetheless have true organs, such as intestines and gonads. This is not true for the most primitive of all animal groups, which hardly fulfill the conventional picture of what an animal really is at all.

Coelenterates are such a group: a vast phylum of animals entirely lacking organs, but getting by with a "tissue-grade" body organization. Tissues are collections of similar cells that are organized to perform a particular function. Organs, such as intestines, are typically composed of several distinctly different tissue types, each performing a separate function and acting in concert to perform the collective functions of the organ. In the case of an intestine, the functions would be digestion of food, absorption of nutrients and water, and the elimination of waste.

Coelenterates (sometimes also known as cnidarians) are the jellyfish that can be such a bane to swimmers; they are also the sea anemones forming small colonies on rocky shorelines. They are also the corals, close relatives of anemones, that form massive and extensive reefs in many regions of the tropical oceans. No matter what form they take, the animal itself is the soul of simplicity, consisting of but two layers of tissues that do all the work—filter-

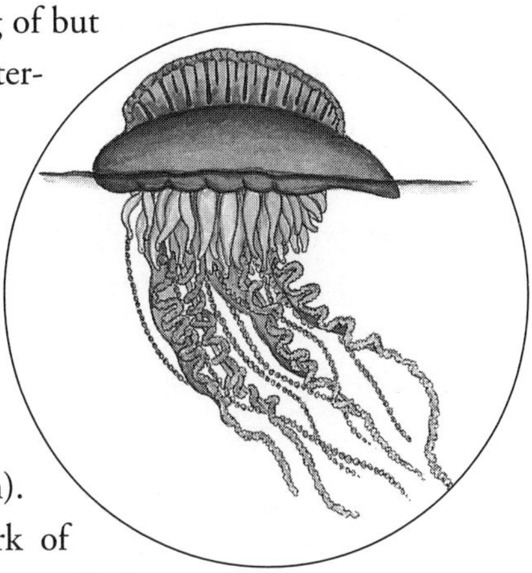

[Figure 34] *A Portuguese man-of-war. Many cnidarians (both polyps, such as corals, and medusae, such as the complex man-of-war) are actually capable of capturing, ingesting and digesting entire fish— making them active carnivores as well as filter-feeding animals.*

ing food particles, absorbing oxygen, and producing gametes (sex cells: eggs and sperm). These layers are united by a loose network of nerve cells that integrate the organism into a unified, functioning whole. Most remarkable of all the cell types of a coelenterate's body is the *nematocyst*, a cell that explodes when stimulated, releasing an arrowlike barb, which are the nettles that afflict—and, in some species, kill—the unwary swimmer.

Many coelenterates, including reef-building corals and some larger kinds of jellyfish (such as the Portuguese man-of-war) form true colonies, meaning that many genetically identical individuals form a single kind of superorganism. [Figure 34] A brain coral, for example, is actually a colony consisting of hundreds of genetically identical and, thus, rather similar organisms. Some coelenterates take the colonial habit a major step further, making up for their lack of specialized tissues and organs in what I find an absolutely fascinating piece of parallel evolution. These colonies of genetically identical individuals somehow manage to develop into radically different adult forms, which fulfill various different functions for the colony as a whole. A man-of-war consists of a number of different body types that form floats, reproductive structures, and stinging tentacles, thereby transforming a colony into what looks deceptively like an individual organism with multiple organs. Yet, it is a colony of genetically identical distinct organisms! When confronted by these differentiated coelenterate colonies, the mind fairly reels at the question, "What is an individual?"

We cannot leave the animal kingdom without dwelling for a moment on the lowly sponges—so primitive that hardly anyone who is not a professional marine biologist even realizes that they are true animals! Even though they lack the true tissues that we find among the coelenterates, sponges do have several different kinds of cells, which are similar in construction to those of higher animals and to certain groups of single-celled, microscopic organisms as well. In their own peculiar way, sponges provide a link between complex multicellular animal life and the hidden realm of microbial life from which we have sprung.

The overall form of a sponge is simple enough—just a body mass enclosing a central cavity; water filters in through small channels in the body wall and is expelled through a large upper opening. The details of the cellular life inside the sponge link these inert creatures to the animal world: so-called collar cells line those side channels, equipped with whiplike tails (*flagellae*) that help the cell trap nutrients from the passing stream. The internal ultramicroscopic structure of these flagellae are identical to the tails of sperm cells in higher animals and to the flagellae of single-celled protists known, unsurprisingly, as *flagellates*. That internal structure is famous as the "9 + 2" ring of nine pairs of internal rods encircling two more pairs in the middle; the rods of the retinas of our eyes also have this structure, one that pervades the microanatomy of all whiplike structures in all animals and animallike single-cell microbes.

Free-living, single-celled flagellates move around like sperm cells, powered by the undulating motions of their flagellae. They are heterotrophs; like true animals, they engulf food, rather than synthesizing sugar from sunlight the way plants and certain photosynthetic microbes do. So do amoebae, though lacking flagellae, they locomote an entirely different way, by extending their protoplasm through pseudopods—thus re-forming the body to move in the desired direction. Amoebalike cells also show up in sponges, as the cells that secrete the spicules—segments of the body skeleton—which are composed of either calcium carbonate, silica (glass), or the more familiar spongy proteinaceous material of the familiar bath sponges.

No question that sponges are true animals, not because sponges prowl around seizing prey, but because the few cell types they have are animallike and also very similar to the most animallike of the single-celled microbial protistans.

CONNECTING DOWN AND LATERALLY: FUNGI VERSUS PROTISTS AS THE CLOSEST RELATIVES OF ANIMALS

We have reached the virtual base of the animal tree (though see Figure 15 for additional information). As we search ever more widely and deeply for our animal roots, the question now is, "What group really does constitute our closest relative?"

We, as animals, share with two other major groups—Kingdoms—the fact that our bodies are constructed of many cells. In our own case, we have *billions* of cells in our bodies, belonging to over 200 different cell types. Fungi and plants (Kingdom Fungi and Kingdom Plantae) are also multicellular, though no fungus or plant comes close to human beings in terms of diversity of cell types.

Thus, we have a choice: We can assume that multicellularity is an evolutionary specialization that links us most closely with either plants or fungi, or we can look—as we just did, when considering what sponges are—at single-celled organisms and contemplate the possibility that the closest relatives of animals may not be other multicellular organisms (i.e., fungi and plants), but rather certain of the vast array of single-celled microbes.

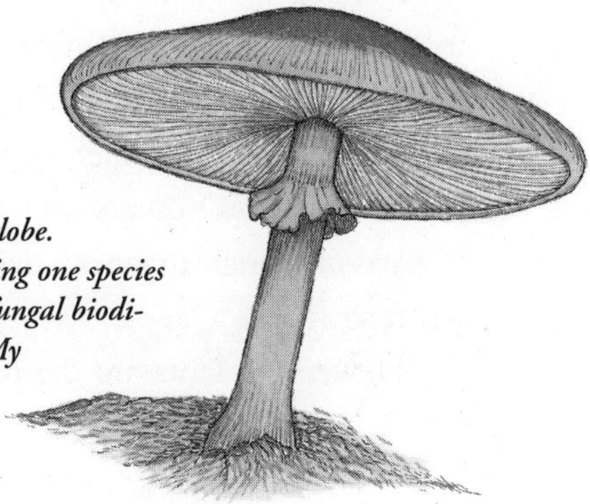

[Figure 35] *Destroying angel mushroom,* Amanda virosa. *Though the potential dangers of poisonous mushrooms are well-known, mushroom collecting for gastronomic purposes remains a passionate interest for people of many different cultures around the globe. No other pursuit underlines the importance of knowing one species from another so dramatically: A little knowledge of fungal biodiversity can greatly enhance your dining experience. My own favorites are the morels that pop out in springtime in the northeastern United States.*

We can also, in a sense, split the difference: We can ask, which of the other two multicellular kingdoms, plant or fungi, is actually more closely related to animals, while still holding out the possibility (as I believe is actually true), that we animals are more closely related to amoebalike and flagellated protists than to either plants or fungi. If we confine our gaze, for the moment, to just plants and fungi, it is legitimate to ask, "Which of these are we closest to?"

It's the fungi. Yep. Fungi—mushrooms, molds, and yeasts being the best known—share more genetic information with us than either they or we do with plants. [Figure 35] Fungi are saprophytic, gaining sustenance directly from dead organisms, meaning plants and animals. They are the recyclers of the planet, without whom literally everything would come to a standstill, because dead but not fully decayed plant material, in particular, would quickly accumulate.

With the exception of the mushrooms, most fungi are inconspicuous, existing as slender filaments, in many cases below the soil surface. Indeed, mushrooms are merely the fruiting reproductive structures of a subsoil organism consisting of a meshwork of such tubes, called *hyphae*. Modest as such organisms seem to be, though, a single soil fungus, *Armillaria*, was recently pronounced "the world's largest organism," weighing in at an estimated 11 tons and covering an area of some 37 acres at Crystal Falls, Michigan. Its size was determined by analyzing the molecular genetics of samples of the fungus and determining that, over that 37-acre patch, all samples were genetically identical and thus all belonged to the same organism! [Figure 36]

Many animals, such as bryozoans and corals, may also reproduce asexually, producing colonies of clones, genetically identical organisms that remain attached to one another. The fungi fit this mold—but so do many true plants—so the current record holder for "world's largest organism" is no longer that Michigan fungus (or its rivals quickly found by competing research teams), but rather a stand of aspen trees in the Rocky Mountains, all connected by runners underground, all genetically identical, and thus all parts of a single organism despite their outward appearance as a stand of separate trees.

Flowering plants are the bastions of terrestrial ecosystems the world over. They photosynthesize, "eating" sunlight in a chemical reaction that

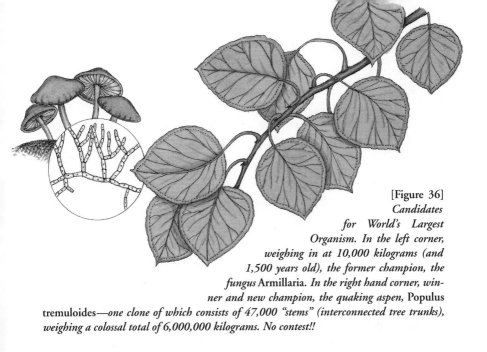

[Figure 36]
*Candidates
for World's Largest
Organism. In the left corner,
weighing in at 10,000 kilograms (and
1,500 years old), the former champion, the
fungus* Armillaria. *In the right hand corner, win-
ner and new champion, the quaking aspen,* Populus
tremuloides—*one clone of which consists of 47,000 "stems" (interconnected tree trunks),
weighing a colossal total of 6,000,000 kilograms. No contest!!*

traps solar energy in the form of simple sugars. The reaction (in its simplest form written as $12H_2O + 6CO_2 \longrightarrow C_6H_{12}O_6 + 6O_2 + 6H_2O$) is me-
diated by the presence of chlorophyll, the catalyst that is responsible for
the green color of plant leaves and thus much of the land surface of Earth.

It is hard to overestimate the profound importance of the chemical
process of photosynthesis: It is the basis for all life, save among a few of the
most ancient bacteria that utilize different pathways to gain energy. Otherwise, all organisms either trap solar energy through photosynthesis, or eat
plants that have done so, or eat organisms that have eaten plants. Photosynthesizers are literally the base of the food chain.

Look at the photosynthesis equation again, and note that plants are
regulating atmospheric carbon, by taking in and converting carbon dioxide—the only way this essential element for all "organic" molecules gets
into the cycle of life, literally extracting it from a common constituent of
the atmosphere. Plants are not just crucial to life, they virtually define it,
control it, run it. By taking in and releasing water, the trees of tropical rain
forests control the water cycle to a large degree; those afternoon thunder-
storms during the rainy season in the Amazon basin reflect water discharge
from the trees through a process known as transpiration.

Perhaps the most graphic demonstration that plants are the very heart
and soul of terrestrial ecosystems is their sensitivity to the physical realm,

[Figure 37]
Sunflower, Helianthus annuus. Sunflowers are members of the vast angiosperm family Compositae. Each flower is literally a composite of many tiny flowers, arrayed in spirals, with each one producing a seed.

particularly to climate change. Plant life is very sensitive to swings in rainfall and seasonal temperatures. For example, as we have already seen, when global temperatures dropped, bands of plant communities—from the low-grass tundra, the northern coniferous forest, the mixed-hardwood forests, and plains—all moved farther south as glaciers descended on Eurasia and North America from the polar regions during four separate periods in the past 1.65 million years. Likewise, when rainfall becomes scarce, moist tropical woodlands give way to more open savanna grasslands, then to drier scrublands, and ultimately to sparsely vegetated true deserts—all a reflection of changing patterns of rainfall. As the plants go, so do the animal species that are adapted to eating them, and *they* take along their predators.

The *angiosperms* or flowering plants are by far the dominant, most numerous members of the modern plant world. [Figure 37] Angiosperms evolved rapidly beginning in the Cretaceous, to a large extent replacing the *cycads* (palmlike trees such as the sago palm, one of the few surviving members of this group) and the *conifers* (pines and their relatives, which go back to the Paleozoic, but are now limited to a scant 550 species). [Figure 38] There are at least 230,000 species of angiosperms. Their name derives from the secret of their success, a seed that houses the embryonic plant and the endosperm to nourish it as it develops and is protected by a tough outer seed coat. Thus, the an-

[Figure 38] *Sago palm,* Encephalortos longifolius. *Cycads, such as this sago palm, and conifers (evergreens) are long-lived in an evolutionary sense, going way back to Paleozoic times. But they can also be incredibly long-lived as individual plants. Some of the giant redwood trees of California go back nearly to the birth of Christ—as is revealed by counting the growth rings on a polished slab of one redwood specimen that has long been on display at the American Museum of Natural History.*

giosperm seed is the plant equivalent to the vertebrate amniote egg, both of which are vital reproductive adaptations conveying great advantages to life on land. The angiosperms include all the familiar wild flowers plus all but a handful of the world's trees, bushes, and grasses. Mosses and ferns are the most conspicuous of the remaining plant phyla—holdovers from the early days of plant evolution that have managed to survive in the post-Cretaceous world dominated, in plant terms, by the angiosperms.

MICROBES We now enter the microbial world. With a few exceptions, these creatures are invisible to the naked eye, and most of us are totally oblivious to their existence. There is almost an inverse law of importance here: The less conspicuous a group of organisms may be, the more it usually turns out to be at the very center of things. It is no exaggeration to say that microbes run Earth. Plants, while crucial to life in the terrestrial realm, are actually outdone by the single-celled algae floating near the surface of the world's oceans, photosynthesizers that together crank out far greater volumes of atmospheric oxygen than all the land-dwelling plants of the world (not too surprising considering that over 70% of the world's surface is covered by the oceans).

There are some 27 phyla of *eukaryotic* (i.e., complex-celled) microbes, sometimes united into a Kingdom Protoctista. As we have already seen, however, some of these more advanced single-celled eukaryotes are clearly allied with either fungi, plants, or animals, implying that protoctistans are not an evolutionarily completely coherent group. Zoomastiginans, for example, are heterotrophs, like animals. Some of them, such as the notorious trypanosomes, cause terrible diseases like African sleeping sickness. We have already encountered others, though, that live in the hindguts of termites, joining with some groups of bacteria and fungi as the only creatures able to process cellulose and thus playing a vital recycling role in Earth's ecosystems.

Amoebae, plus shelled amoebalike forms such as marine Radiolaria (whose delicate glassy skeletons are marvels of beautiful construction) and foraminiferans, are also heterotrophs and clearly very much like animals. On the other hand, Phylum Chlorophyta—the green algae—are photosynthetic and obviously very closely related to true plants. Indeed, several lineages of chlorophytes have become multicellular (though they lack the tissues and organ systems of the multicellular true plants). *Volvox* is an amazing chlorophyte, forming colonial globes whose cells cooperate in rhythmic synchronicity, enabling the colony to move around in concerted fashion.

Other important protoctists include the euglenophytes—fascinating combinations of plantlike and animallike features. Euglenids have flagellae, which they use to propel themselves through their aqueous environments. Most of them are photosynthetic, but they can also feed like animals, which perhaps places these creatures near

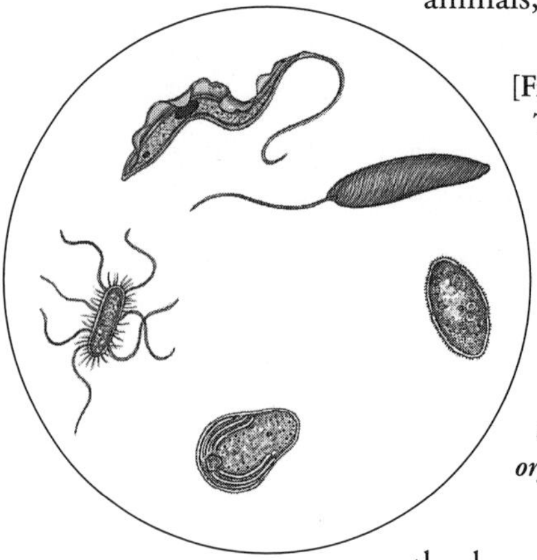

[Figure 39] *Microbial life. Clockwise from upper left:* Trypanosoma brucei *(a protoctist),* Escherichia coli *(intestinal bacterium),* Nitrobacter winogradskyi *(a chemoautropic bacterium),* Methanobacterium ruminatium *(a methanogenic bacterium), and* Euglena spiroggra *(a protoctist). The microbial world presents the last great frontier of ignorance in biodiversity; new tools and techniques of molecular biology and electron microscopy have recently begun to reveal this realm of the tiny, but ecologically vital, organisms.*

the base of the higher protoctists before the grand three-way split into fungal, plant, and animal directions.

Easily the most arresting of all eukaryotic microbes—to my mind, at least—are the cellular slime molds of the Phylum Acrasiomycota and the very similar, sexually reproducing plasmodial slime molds of the Phylum Myxomycota. What fantastic creatures they are: Individual single-celled heterotrophic amoebas living in freshwater, damp soils, and rotten logs, and periodically aggregating to form a single organic structure often visible to the naked eye. In the case of the cellular slime molds, this structure is a reproductive body that releases spores that germinate into a new batch of amoebas, which eventually all come together to form yet another spore-producing individual as the reproductive cycle continues. The amoebas move like animals, absorb nutrients like fungi, and form spores like plants. Fantastic!

We delve even deeper. Biologist Lynn Margulis has proposed that the complexly celled eukaryotes are the outcome of a symbiotic event in the early stages of evolutionary history, when representatives of two different lineages of simple-celled prokaryotes—bacteria—fused to form the earliest complex-celled form. For roughly the first 2 billion years of life on Earth, bacteria were literally the only organisms around. Earth was theirs.

What is not immediately obvious is that Earth still belongs to bacteria, despite the ubiquity and far more obvious presence of us eukaryotes. [Figure 39] There are nine major groups (phyla) of bacteria still very much with us. Some, such as the blue-green algae, are photosynthesizers. Others, such as spirochetes, work alongside protoctistans (such as zoomastiginans) in the hindguts of termites, breaking down cellulose. Other spirochetes are not nearly as friendly to our interests, causing syphilis, Lyme disease, and other unpleasant human afflictions.

Escherichia coli, which lives by the billions in the intestine of every living human being, is a decidedly mixed blessing. Although its presence is vital to the normal workings of our guts, virulent strains quickly produce diarrhea and fever; a related bacterium causes cholera, Rocky Mountain spotted fever, gonorrhea, and meningitis. Some particularly depressing species convert wine to vinegar. Yet, of the group as a whole—Phylum Omnibacteria—Lynn Margulis (premier champion of the microbial world) has written that "it is not too extreme to say that most life on Earth takes the form of facultatively anaerobic . . . unicellular rod-shaped bacteria"—in other words, these guys.

"Facultatively anaerobic" means that these bacteria can switch at will from living in oxygen-free to oxygen-rich environments. Most other organisms either must have, or cannot survive in, oxygenated environments. Such physiological flexibility is a mere tip of the iceberg when it comes to bacterial metabolic pathways. Consider the *chemoautotrophs*, bacteria living without sunlight or preformed organic compounds. They make do entirely on air, salts, water, and an inorganic energy source such as sulfur compounds. Called "metabolic virtuosos" by Margulis, chemoautotrophs are crucial to the cycling of nitrogen, carbon, and sulfur in the world's ecosystems.

Another small group of bacteria has considerable capacity for breaking down organic compounds, including hydrocarbons. When the *Exxon Valdez* went aground in Alaska's Prince William Sound, spilling a vast quantity of crude oil into the icy waters, much of the cleanup came through the work of the lowly bacterium Pseudomonas, which is able to eat the crude and thus mitigate, to some considerable degree, the ecological damage wrought by the spill.

We are near the roots of all life. The two most primitive of the living groups of bacteria constitute the Archaebacteria and are holdovers from the earliest days of life on Earth. Methanogens, which are poisoned on contact with oxygen (the earliest atmosphere of Earth was virtually oxygen-free), as befits holdovers of the most primitive life forms, cannot use preexisting organic compounds (such as sugars, proteins, or carbohydrates) as sources of either carbon or energy. Rather, they derive both by combining carbon dioxide (CO_2) and hydrogen to form methane (swamp gas). They are estimated to produce 2 billion tons of methane a year, 30% of which comes from the guts of cows, where some of these methanogens live!

The other archaebacterial group is likewise physiologically unusual and consists of organisms capable of living in the most extreme environments: highly concentrated brine pools (death to all other organisms) and hot springs. Their habitats show just how ubiquitous life is on this planet, and that the most extreme conditions are just right—not for the creatures produced relatively recently after billions of years of evolution, but for the direct descendants of life's very first representatives on the primordial Earth.

There you have it—a quick overview of life's rich evolutionary diversity, a scorecard of the major players in the game of life. What, though, of viruses? Aren't they alive as well? In a word, No. Viruses have RNA but are utterly unable to reproduce on their own. Though much smaller and simpler than bacteria (suggesting that they are an even more primitive form of life), viruses are parasites, utterly dependent on the DNA of living creatures, whose cells they invade, take control of, and use to replicate themselves. Thus, viruses—whatever they are—must have appeared *after* bacteria and eukaryotes. Fantastic as it may seem, I think there is much to the suggestion (yet another one from microbiologist Lynn Margulis) that viruses are escaped snippets of the genetic code of other organisms. Rather than forming a natural group, viruses—such as those that cause polio, AIDS, or flu in humans—are perhaps more closely related to us than they are to, say, the tobacco mosaic virus. Important as they nonetheless are to the well-being—or lack thereof—of the world's true organisms, viruses stand to the side in the spectrum of life.

Chapter 4 ECOSYSTEM PANORAMA

We've met the players, and now it is time to visit the major ecosystems, the actual arenas where the game of life—the essentials of energy gathering, nutrient cycling, and reproduction—is actually played out. Ecosystems, recall, are linked spatially, as energy flows between local ecosystems, forming increasingly larger-scale regional systems. Ultimately, the entire surface of Earth is covered with an interlinked array of life, forming the Biosphere. If evolution produces the players in the game of life, it is the world's interconnected array of ecosystems in which the game of life is played. ❦ *It is a singular fact that a traveler, starting at either of the geographic poles, will find the number of different species populating local ecosystems increasing as the equator is approached. For animal life, the ice floes* of the Arctic Ocean (there is no land at the North Pole) support polar bears, living off fish and seals. The arctic islands and far northern expanses of the Eurasian and North American continents mark an increase in animal diversity, with caribou and musk oxen surviving on the relatively few species of tundra grasses and other vegetation that spring up each short summer. In addition, shorebirds, gulls, and waterfowl utilize the tundra to breed during the summer months, as do several species of songbirds (such as snow buntings and longspurs), while rough-legged hawks dine on arctic hare, as well as on lemmings and related rodents.

Many of these species occur right around the circumference of the far north: The wolves of Siberia belong to the same species as those of Alaska and Manitoba (*Canis lupus*); likewise, caribou, musk oxen, and rough-legged hawks are all members of far-flung species that, in some cases, completely ring the northern sector of the planet. Though there seem to be relatively few species comprising the ecosystems of the polar regions, one other fact jumps out to the traveler's eye: There are usually many individuals of each of those species. Vast herds of caribou (reindeer) were at least primordially common, and the pattern applies even to the carnivores: Relatively few species (wolves and arctic foxes in the dog family, sharing the carnivore role with grizzly and polar bears), but fairly large numbers.

Jump for a moment to any tropical rain forest sitting astride the equator. Though birds and mammals are often difficult to detect, patient study typically reveals hundreds if not thousands of birds and mammal species in places like the Congo Basin in Africa or the Amazon Basin of South America, showing much greater diversity than in the higher latitudes. Although the dense forest cover is a profound impediment to spotting birds and mammals in this setting (as any novice birder, eager to spot tropical species, will attest), tropical species can be elusive for another reason. With some notable exceptions, tropical species cover smaller areas and live in smaller populations than their kin of the higher latitudes. If we compare open tropical grasslands with tundras, we do see vast herds of wildebeest still trooping across African savannas, but a comparison of the number of herbivorous mammal species in the arctic with those of Africa (recall the great diversity of antelopes in Botswana's Okavango Delta!) amply confirms the general pattern. There are far more species of large herbivorous mammals in any given habitat in Africa than in any part of the northern tundra, but the population sizes of the relatively few north-ern species are generally far greater than the average population sizes of the Tropics.

Why are tropical ecosystems so much more diverse than their polar counterparts? Organisms exposed to wide daily and seasonal fluctuations in temperature so typical of the higher latitudes must have adaptable phys-iologies; they must be able to derive sustenance from a wide variety of foodstuffs as well. Photosynthesis shuts down in the higher latitudes in winter, as the juices inside plants would simply freeze. With plants dor-mant, insects disappear. Most die and leave their eggs behind. Other in-sects become dormant or—as in one very celebrated instance of the monarch butterfly—they migrate the way many bird species do. Some bird species, such as North American yellow-rumped warblers, drop south a bit, but don't join their fellows for a winter in the Tropics, switching from insects to seeds to tide themselves over the colder months. Some mammals hibernate, but others scratch out an existence, foraging beneath the snow for remnants of last summer's vegetation. You have to be adaptable, flexible and able to take strong swings in temperature and to exploit a wide variety of energy sources to survive year-round. Adaptability becomes more im-portant in higher latitudes, north or south.

Hardiness and adaptability—these are the watchwords for the full-time denizens of high-latitude terrestrial ecosystems (all but the nearest-shore marine systems are more buffered and less subject to daily and seasonal oscillations). Ecologists call such species *broad-niched*. It stands to reason that such broadly adapted species are able to thrive in a wide range of environments and, for that reason, tend to occur over large expanses of territory.

In contrast, in the Tropics, species live in far more predictable climes and have far more day-in, day-out, season-to-season reliable food re-sources. That is why species in the Tropics tend to be more narrowly adapted than their equivalents in higher latitudes. In an evolutionary sense, species in the Tropics focus on microhabitats so that the overall pat-tern is one of specialization. Tropical species are far more likely to divide up the ecological resources, specializing on one particular food resource, or, in the case of plants, adjusted to soil with slightly different proportions of silt, sand, and organic materials, or slightly different amounts of annual rainfall. Specialized as they are, species in the Tropics, for the most part, do not occur over comparably vast areas, as is more typical of species farther away from the equator. The tendency toward specialization ultimately means that more species can be packed into local ecosystems in the Tropics than could ever be the case in the higher latitudes, where species tend to generalize, overlap, and compete for resources.

Thus, new species have a better chance of survival in the Tropics—at least for a while. Specialization is, however, a double-edged sword, and tropical species are notorious sitting ducks, prime targets for extinction. For example, when global climate cools, the major habitat zones tend to converge toward the equator, and many species of, say, the tundra have no difficulty continuing to find suitable, familiar habitat. Tropical species have nowhere to go, and when *their* habitat changes, they are prone to extinction.

There is another element to the story, one that is especially crucial to the modern extinction crisis engulfing the world's ecosystems and species. The element in common between mass extinctions of the remote geologic past and the current dilemma is *habitat destruction*, which occurs these days through the hand of man, as we continue to convert terrestrial ecosystems to agricultural soils; chop down forests for pulp, building materials, and firewood; and continue the process of paving over the countryside for commercial and residential use. The process is most advanced in the higher-latitude, technologically advanced, so-called developed nations such as the United States.

Yet relatively few species have so far completely disappeared in these regions. Why? Because they are so widespread and adaptable, most of the species of the temperate regions, though cut back dramatically in numbers, are nonetheless not yet on the brink of extinction. The fossil record shows us why: The best predictor of survival of a mass extinction event turns out to be the size of the region a species occupies. The greater the area, the less likely a species will become extinct. This translates directly into what we see going on around us today: Despite the great transformations of habitat that have already occurred in industrialized North America and Western Europe, few species have become extinct.

As developing nations of the Tropics begin to follow suit, the fear is that we are losing thousands of species each year. Why? Because species are much more narrowly distributed in the Tropics. The point is critical, because tropical countries have, logically enough, pointed to the habitat conversion of developed countries in the temperate latitudes and have, in essence, complained of hypocrisy on the part of conservationists from the developed countries who point with mounting horror to the destruction of the world's tropical rain forests. "You have already done it, and now you are trying to deny us the development of our own resources" is the not-unreasonable cry. Unfortunately, it is true that the Tropics hold many more species and more specialized species in much narrower distribution than do the temperate zones. Now, as in the past, tropical species are virtual sitting ducks for extinction through habitat conversion and degradation. Somehow, the biology of the matter must enter into the dialogue before it is too late.

THE WORLD'S ECOSYSTEMS: A PERSONAL TOUR Let's take a trip from the arctic ice floes and

work south through the tundra and eventually to the equator. Something of the structure, pulse, and interactivity among the components of typical ecosystems emerged in our consideration of the Kalahari dry grasslands and Okavango Swamp wetlands in chapter 1. On a smaller scale, the complexities of mini-ecosystems were revealed in the inner workings of termite mounds, a fixture of both the Kalahari and Okavango systems. What follows are necessarily briefer profiles of the other major categories of Earth's ecosystems, with emphasis on some of their characteristic players, their component species.

As the ecosystem tour progresses, one thing to look for is how evolution time and again produces similar organisms—hence similar ecosystems—in comparable habitats (such as deserts or wetlands). Often, distantly related organisms will appear to have the same basic evolutionary adaptations equipping them for life in their native environments, a phenomenon known as *convergent evolution.* The high latitudes of the northern hemisphere, for example, have alcids—small, generally black-and-white birds, which are the less familiar analogues of the penguins of the southern hemisphere. Alcids include puffins, guillemots, and razorbills, as well as the penguinlike great auk, the only known species of this group not able to fly. When archeologists discovered cave paintings in a submerged Mediterranean grotto in the mid-1990s, one of the birds depicted was proclaimed to be a penguin—the first of its kind ever documented in Europe—until it was realized that the bird was really a great auk! Great auks became extinct in 1844, victims, not of habitat destruction, but of their willingness to march up ship boardwalks, unwittingly joining the ship's larder. They were victims of overhunting.

Tundras are treeless expanses of the higher latitudes. Vegetation consists of low grasses and bushes; my sole personal encounter with tundra so far has been on St. Paul Island in the Pribilofs, focal point of our earlier discussion (chap. 3) on the dire threat to modern marine fisheries. There, arctic foxes romp in exuberant abandon over two distinct sorts of vegetation. The most common covering is a species of wild celery, which grows to knee height. On the western part of the island, a very short grass predominates; in both regimes, summer wildflowers are also plentiful. The theme is striking: relatively few different species, but many individuals of each of them. Yet, on St. Paul, there is ample evidence that even near the Arctic Circle, some form of specialization still goes on: On a narrow spit of land less than 90 meters wide, I had no trouble spotting rosy finches, snow buntings, and Lapland longspurs—three small songbirds of the open tundra, the males singing their hearts out. Yet, at the edge of the cliff only a few meters away, puffins, kittiwakes, and cormorants—all focused on the sea—wheeled just off the cliff face: I never once saw any of these seabirds cross that spit of land, or a finch or longspur actually fly out over the edge of those prodigious cliffs.

Below the tundra lies the boreal forest, a thick expanse that, in North America, crosses northern Canada, dipping down into the United States just into northern New England and the Adirondack Mountains of New York State. I have spent more time in the Adirondacks than in any other near-pristine natural environment. These northern forests harbor many different species of hardwoods—meaning angiosperm trees such as striped maples, several species of birch, and beeches (whose berries are much beloved by the local black bears). But it is the conifers that dominate—in the case of the Adirondacks, especially hemlock—and at full, undisturbed maturity, white pines of tremendous height and girth. It is almost as if the angiosperms pushed the gymnosperms (conifers) aside with their explosive evolutionary burst beginning in the Cretaceous, leaving the harder climes as bastions for conifer survival.

In the Adirondacks it is relatively easy to observe the process of "succession," in which pioneer species invade an area and literally take over, only to yield in a few years to other species. As the forest thickens and grows ever upward, each tree competes with those around it for sunlight, with eventually only the tallest, longest-lived trees left standing. The process is extremely patchy—meaning that one spot will be degraded, leaving adjacent areas untouched and thus at a higher level of succession, while the degraded patch must start all over again.

Say, for example, a fire sweeps over the forest or, as happened in the summer of 1995, a violent wind storm cuts a wide destructive path across the central Adirondacks. In the latter case (and wherever humans do not clear away the fallen logs), insects, fungi, and microbes must decompose the twisted mat of dead trees before succession can begin. In the case of fire, succession resumes much more quickly, typically with aspens, striped maples (if the soil is sufficiently moist), and a few other weedy species quickly becoming established. Soon silver and white birch become established, possibly crowding out the quick-growing but short-lived pioneer species. Only later come the slower-growing—but ultimately dominant—species such as white pine and hemlock.

Though succession is a *temporal* concept (meaning simply it occurs in one place over time), the patchy nature of both natural and human-engendered habitat degradation means that all stages of succession can be found cheek-by-jowl. The most striking example that I have yet seen was also in the boreal forest, but this time in Glacier Bay, in southeastern Alaska. Today, Glacier Bay lies wholly within the confines of Glacier Bay National Park. Visitors are treated to spectacular vistas of cliff faces and glaciers—literally rivers of ice—coming down and spalling off icebergs at water's edge. A trained eye—or, as in my case, one guided by an expert Glacier Bay Park Ranger—sees something else. At park headquarters, near the mouth of the bay (where a humpback whale was idly cruising and snorting one morning when I visited), there is a full-blown, typical spruce-dominated northern forest, replete with blue grouse, hairy woodpeckers, and a host of other animal species. As the boat proceeds slowly up the full extent of the Bay, that forest gradually begins to disappear; birches and aspens begin to predominate, and, as one cruises still farther north, eventually all trees disappear, and the only plant life is small bushes and eventually sparse grasses. By the time the head of the bay comes into view, there is nary a plant to be seen.

Here is succession with a vengeance: 200 years ago, Glacier Bay was completely filled with an enormous glacier. Then, a remarkable thing happened. The glacier was growing so fast that it over ran its own *terminal moraine*—a ridge of sand and rock that forms a wall in front of an advancing glacier as it bulldozes and scours the landscape. The mouths of all of Alaska's fjords (each one carved out by its own glacier) are very shallow, recording the furthest extent of the terminal moraines that front all glaciers. With the glacier's morainal wall suddenly gone, the sea attacked, eroding this mammoth glacier some 105 kilometers in the past 200 years!

Rapid as this retreat was, however, the ice left the side of the valley down near the mouth 200 years before it was finally wrested from the banks of the fjord 105 kilometers up the Bay. Plants had just that much longer to invade, establishing soils by eroding rock particles, solidifying them with roots, and, upon their deaths, adding to the organic content, and producing soils in which larger species could take root.

Just outside Glacier Bay lies Tongass National Forest—a monumental expanse of mature, Sitka spruce-dominated forest, home to the famous Alaskan brown bear, wolves, deer, loons, and the more-often-heard-than-seen varied thrush, whose melodious warble symbolizes the very essence of the deep, wet forest. It is a gut-wrenching shock to sail by the magnificent forest-clad islands of southeastern Alaska and suddenly see a clear-cut, as if a giant had taken a razor and simply shaved a swathe across the landscape. The issues here are complex, and because the islands themselves are largely uninhabited, tend to revolve around the sale of forest timbering rights, especially to foreign interests—often, it is alleged, at fire-sale prices.

More difficult to pass judgment on—at least for me—are forestry and other aspects of land use in places better known to me, such as the Adirondack Mountains. As we have already seen in our consideration of the Botswana microcosm (chap. 1), conservation efforts cannot ever work if they are enforced by outsiders (regional or national government, or foreigners, such as well-meaning conservation-minded persons living far from a threatened area), *and* if they are seen as inimical to the economic interests of local peoples living in and around the area. Poverty stalks Adirondack villages, especially in winter, and resentment runs deep when downstaters and politicians dictate what locals can and cannot do with "their" forest. Yet, it was the wise foresight of New York politicians and public-spirited citizens, who, in 1892, established as "forever wild" the part-public, part privately held 2.4-million-hectare bundle of land that remains by far the largest protected expanse east of the Mississippi. Were it not for that interference, the Adirondacks would almost certainly not be as (relatively) healthy an area as it now is. The real point here is the complexity of the issue and that somehow—and every situation is different—the economic needs of local peoples must become confluent rather than competitive with the desire to stem the tide of ecosystem destruction.

OF PRAIRIES AND WOODLANDS On February 15, 1995, the *New York Times* carried an intriguing report on the status of the primeval prairie of the great American Midwest. When European settlers first arrived and finally penetrated the Appalachian Mountains, they encountered extensive woodlands—temperate forests dominated by such angiosperm hardwood trees as sassafras, hickory, oak, maple, and ash, stretching beyond what is today Ohio and Indiana. Across the Mississippi stretched vast plains of grass, dotted here and there by prairie dog towns, small herds of pronghorn antelope, and occasional vast herds of bison, American buffalo.

Those extensive eastern woodlands are now largely gone, felled by the ax to yield wood for building, logs for fires, and above all to clear the land for agriculture. The prairies have fared little better: The endless grasslands have long since been transformed into "amber waves of grain," as agricultural monocultures have all but completely replaced the natural prairie ecosystems. The tremendous, devastating Dust Bowl of the early 1930s, superimposed on an economy already reeling under the Great Depression, robbed these western lands of much of its hard-accumulated top soil—a disaster that resulted from a combination of drought conditions and unsophisticated farming practices. Loss of top soil is one of the most serious environmental threats faced by humanity as we prepare to enter the 21st century.

It's hard to believe that all but tiny remnants of the prairie are gone, but the capacity of natural ecosystems to restore themselves must never be underrated. The *New York Times* article focused attention on some weedy lots sandwiched between two railroad lines on the outskirts of Chicago. The land was "useless": Though once farmed, the advent of the railroad had rendered the land unattractive. Quietly, unnoticed by all save a few hikers and nature lovers, the prairie began, ever so modestly, to regain its own. Most spectacular in the spring, when a riot of colors from blossoming native wildflowers would catch the eye, to most passersby these places were simply weedy lots. [Figure 40]

Most vacant lots in modern America are indeed choked with alien grass and other plant species, many of which were introduced from Europe during the years of colonial settlement. These lots outside Chicago are different: They are staffed by members of native grass and flower species. Species act as reservoirs of genetic information, and, whenever conditions appear that resemble primordial habitat, the old dance of ecological succession of the Midwestern prairie will begin, provided that colonists (seeds, in the case of plants) can find their way there. Precisely because high-latitude species tend to be spread over such broad areas, even the thorough conversion of these midlands into the rich breadbasket of the nation failed to destroy every last acre of primordial prairie. Enough was left so that no species of prairie plant is known to have become extinct. They linger, in small outposts such as the little prairie microcosm between the tracks in the Chicago suburbs, ready to send out emissaries to reclaim the terrain should the combined agricultural, industrial, and urban enterprise of the great American interior ever relinquish its grip.

It is good to know that, given just half a chance—as long as habitat conversion hasn't pushed the species that comprise a given ecosystem completely to extinction—nature is able to rebound from prodigious onslaughts. It should give heart to those who would see the Sixth Extinction thwarted, but who fear that it might already be too late. Although substantial numbers of locals—perhaps even the majority—interviewed by the *New York Times* about that little bit of Chicago prairie evinced no interest (some thought the lots an eyesore, best converted to—what else?—another shopping mall), many seemed intrigued and even thrilled to know that the prairie, while muted, is still with us, with the potential for its return very much still there.

Elsewhere, especially west of the Mississippi, prairie restitution programs have been gaining momentum. Perhaps not surprisingly, it turns out that there are several distinct types of prairie to be found between the Mississippi and the Rocky Mountains. Through the efforts of state and federal government programs and private organizations (such as *The Nature Conservancy*), the number of acres of prairie land has increased. The success of these programs has been helped by recent realization that fires are natural to maintaining the prairie system (by continually resetting the successional stages and thus providing continued opportunity for all prairie species to thrive) and by programs that encourage natural patterns of grazing by the big herbivores (especially bison). Human management, in areas surrounded by human-disturbed ecosystems, is unfortunately essential but can be successful when close attention is paid to the natural dynamics and underlying ecological processes inherent in the system.

Thus, tallgrass prairies—dominated by two species of bluestem grass—are making their way back in parts of Kansas and Oklahoma. I have visited one such sector just outside Wichita, Kansas, and though the grass is not yet tall enough to conceal a mounted horseman, there are patches that dramatically reveal just how tall those species can grow. Tallgrass prairie contrasts mightily with the shortgrass prairie found just east of the Rocky Mountain Front Range in Colorado. Here, the grass is more like a putting green—a soft, almost velvety texture that, to the untutored eye, looks as if it had been cropped by sheep. Standing in the shortgrass, amid prairie dog towns that are also home to rattlesnakes and burrowing owls, is a very different experience from the primordial Kansas tallgrass prairie. Both are a part of the North American ecosystem heritage—each in its way fragile and drastically cut back, but still very much with us—with the potential for reexpansion should we simply allow them to do so.

Grasslands are ubiquitous. In the northern hemisphere, the interiors of the North American and Eurasian continents consist largely of prairies and steppe lands. In contrast, the two main continental masses of the Tropics, Africa and South America, have grasslands lying peripherally to the more familiar tropical rain forests. In South America, the rain forests of the drainage systems of the vast Orinoco and Amazon Rivers are *entirely* ringed by grasslands, which are home in some cases to as many species of certain groups as are the notoriously species-rich rain forests. A recent

EASTERN BLUE BIRD

COOPER'S HAWK

MEADOW LARK

PRAIRIE WHITE-FRINGED ORCHID

SWEET BLACK-EYED SUSAN

FIRE-PINK PET

GREAT SPANGLED FRITILLARY

[Figure 40] *Weedy lot outside Chicago—with the Sears Tower looming in the distance. For any given kind of ecosystem—such as the prairies of the North American midwest— those that have relatively more of their original species present tend to thrive better: soils retain nutrients, more new plant and animal individuals appear each year, and the system as a whole is better equipped to withstand environmental stress.*

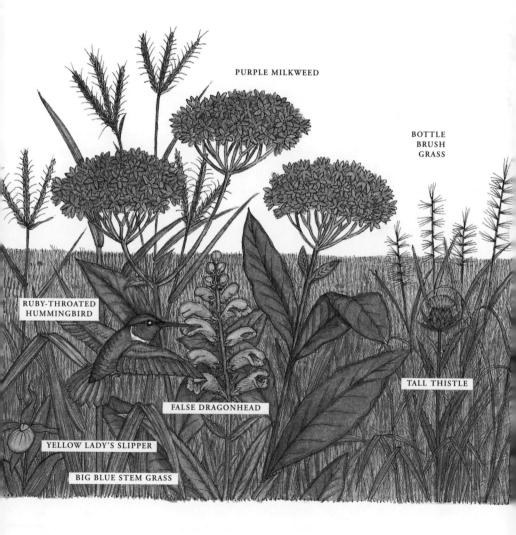

PURPLE MILKWEED

BOTTLE
BRUSH
GRASS

RUBY-THROATED
HUMMINGBIRD

TALL THISTLE

FALSE DRAGONHEAD

YELLOW LADY'S SLIPPER

BIG BLUE STEM GRASS

study, for example, actually revealed *more* species of mammals living in the grasslands around the Amazon-Orinoco rain forest than have so far been discovered within the forests themselves. The situation is similar in Africa, where the rain forests of the Congo and adjacent basins are ringed to the north by grasslands (much of which are now undergoing a process of desertification—victims of ongoing drought and the southward expansion of the Sahara). To the east are the famous plains of East Africa—usually called savannas—dotted here and there with acacia and other trees in small woodland stands. And we have already encountered the southern expanse of grassland—in the Kalahari and wetter climes of the Okavango Delta of Botswana—lying below the classical African rain forests.

Prairies and grasslands grade, sometimes imperceptibly, into drier and more sparsely vegetated lands, often simply called "deserts," even though that term is usually reserved for regions receiving less than 100 millimeters of rain a year. Arid lands—with their often very hot days, and often surprisingly cold nights (I have seen ice on the seats of tractor-drawn carriages parked at dawn on the banks of the Nile, on the edge of the Egyptian desert)—call for complex adaptations, especially to handle the short supply of water. Some species are seasonal, only coming in when surface water is temporarily available. Year-rounders resort to a variety of expediencies: Western hemisphere cacti are famous for their water retention capability, a capacity not lost on some animal species (including humans) that have learned to tap such water stores. Though entirely unrelated, African succulents living in arid regions bear an amazing resemblance, and a very similar physiology, to cacti—a profound case of convergent evolution. Roaming the arid wastes of the Namib desert (west of the Kalahari, primarily in Namibia), I have encountered vistas hauntingly evocative of the Sonoran desert regions of the southwestern United States and Mexico. Perhaps even more astonishing has been the evolution of the ability of some animals to do entirely without the presence of free-standing water, deriving all they need from their food, be it the juices of plants or the blood of other animals. For example, the gemsbok, a large and strikingly marked African antelope of the Kalahari, does just fine without water for much of the year.

Deserts are for the most part considered wastelands by agriculturally based human societies—a fortunate state of affairs, for the most part, as desert ecosystems are extremely fragile and seemingly less able to bear the brunt of human disturbance than are ecosystems richer in that most precious of all substances, water. Yet, with successful efforts "to make the desert bloom" (perhaps most notably in Israel and the southwestern United States), not even deserts are immune to human conversion. Reclaiming marginal land for agriculture is, of course, a notable human achievement. It also most dramatically highlights perhaps the gravest threat to human existence as we know it: the shrinking per capita availability of safe, clean water for drinking, cooking, washing, and agriculture. Though irrigation through tapping underground aquifers continues, the aquifers are rapidly becoming depleted. That leaves only surface waters of lakes and especially rivers available for human consumption. As population increases, and the demand for water in cities begins to compete more and more with both agricultural and suburban residential needs, political collisions over mounting shortages are on the rise.

Tropical rain forests—riots of living diversity and icons of the Sixth Extinction—are the last big segment of terrestrial ecosystems to consider. They provide, as well, a natural segue into freshwater ecosystems per se. As their very name implies, water is as dominant a feature of tropical rain forests as is the lush vegetation. Indeed, virtually the *only* way a casual visitor can see the Amazonian rain forest is from the water, which provides the only easily accessible "roadways"—hence clearings—in the dense tropical forest.

The Amazon is navigable up to (and well beyond) Manaus—1,600 kilometers upstream, and once the resplendent capital of the rubber industry. (The rubber business collapsed suddenly in the first decade of the twentieth century, when Malaysian rubber trees, born of seeds smuggled from the Amazon, matured, and their rubber hit the market at much lower prices). But though large ships regularly ply the main channels of the Amazon, don't expect to see much from the deck. The river is so wide—*kilometers* wide—that no details of the jungle are visible. Indeed, even the shoreline itself can be invisible when farmers are burning off cleared lands, hoping to recharge the nutrient-poor tropical soils for yet another growing season. The way to see the Amazon rain forest is to sail up a large tributary and park at the mouth of one of its tributaries, hop into a small motor boat, and find a side channel (a third-order tributary), which may be only 3 or 6 meters wide. Then and only then does the trop-

ical rain forest come in up close and personal—and the nature of its varied local ecosystems emerge.

It is entirely possible to sit, as I have done several times, in one of a
convoy of several zodiacs (small motorized rubber crafts much beloved in
ecotourism circles) and see almost entirely different birds and mammals
than those seen by passengers in the zodiacs in front of and behind you.
Everyone is likely to see the three-toed sloth high in a *Cecropia* tree, or one
of a troop of red howler monkeys vocalizing like mad but otherwise staying pretty still at their eating spot high in the trees. Hoatzins—large, rather
strange-looking cuckoo relatives—might be hanging out by the water's
edge, and likewise visible to all who pass by. But not so the antbirds,
wrens, tanagers, and other smaller species that flit from the shadows and
disconcertingly quickly disappear back into the foliage. Here I experienced
just the opposite of the scene on the mini-tundra of St. Paul's Island in the
Pribilofs of the Bering Sea: There, where visibility was impeded only by
rain and fog banks, species were few, but individuals were numerous. In
the Amazon, it's just the reverse; *knowing* in advanced that this pattern
ought to be the case is no substitute for the astonishment (and sometimes
disappointment) of hearing—back on the mother ship—that the zodiac
just ahead of you saw a trogon or a jacamar totally missed by members of
your own zodiac. Yet the aracari and guira tanager you saw totally eluded
people in front and in back of you! Here is the pattern we have already met
in the abstract come home with a vengeance: Not only the dense foliage,
but also their relatively small population sizes, obscures the birds and
mammals within the tropical rain forest. Because of this, the probability of
seeing a high percentage of all the species living in any one place is much
lower than when traveling through ecosystems in the higher latitudes.

Satellite photography reveals the inexorable march of rain forest loss.
Especially striking to me is a series of maps prepared and published by biologists Glen M. Green and Robert Sussmann, showing the loss through
human deforestation of 50% (3.8 of 7.6 million hectares) of the eastern
rain forest in Madagascar between 1950 and 1985. It is estimated that,
prior to the arrival of humans in Madagascar a scant 2,000 years ago, the
eastern rain forest was originally 11.2 million hectares. Madagascar, isolated from the African mainland since the fragmentation of the supercontinent Gondwana beginning some 160 million years ago, has what is

known in biological circles as a very high level of *endemism*, meaning that organisms living there are to be found nowhere else in the world.

It is no very great surprise that an island has its own unique species of an otherwise very common group; for example, the Madagascar malachite kingfisher looks a lot like the mainland African species. Madagascar, however, has entire families and subfamilies of birds found absolutely nowhere else: The vangids are a diverse lot related to shrikes, including a sickle-billed species with a long, thin, strongly curved bill, and the helmet bird with a massive, short, stubby crushing bill. Similar adaptations can be found elsewhere, for example, among the honeycreepers, a family of birds endemic to the Hawaiian Islands. But the vangids of Madagascar are an entirely separate evolutionary radiation, and all but three of their nine species are adapting poorly to the degraded habitats that expand as the forests are cut and burned.

Sometimes it seems that nothing indeed can stem the tide of habitat destruction. The Malagasy peoples of Madagascar are, for the most part, desperately poor and struggle to eke out a living largely through a rice-based agriculture. We had thought, on arriving there, that surely at least the steep escarpment on the eastern side of the island—seemingly useless even for terrace-based rice culture—would be spared. At the very first site we visited, we saw walls of flames—not just the usual burning of last year's agricultural stubble, but actually burning down the forest itself to clear the land. If ever there were a place where the economic needs of the local citizenry must be melded with the conservation of the rain forest, it is Madagascar. Failing that, we stand to lose in a very short time a truly unique biological heritage: A land with 9 species of baobab trees (continental Africa has but 1); a land with 32 species of lemurs that occur nowhere else (save 2 species in the neighboring Comoro islands); a land of huge chameleons and endemic birds. And, indeed, a land with traditional cultures—a unique blend of Africa and Asia—that are every bit as much at risk as its rare animals and plants.

FRESHWATERS OF THE WORLD The great
cycle of waters rising up into the atmosphere from the ocean surface, swept by winds in storm systems across islands and continents, and then released

as rain, snow, hail, or sleet but always destined to return to the oceans, is an essential feature of this planet and as vital to life as the ongoing production of oxygen. To be sure, water evaporates from lakes and is released by trees to fall as rain and snow. Ultimately all waters sitting on and flowing over the land masses make their way to the sea—save just a little that manages to seep down and enter the aquifers. Even groundwater—the water table of surface soils—intersects the surface and drains away, as anyone who has a shallow well knows all too well as summer deepens and the rains become fewer and farther between.

Even lakes—so permanent looking—are ephemeral. The Great Lakes that for the most part straddle the U.S.-Canadian border were formed as thick sheets of ice scoured out huge basins during the past Ice Age. The Niagara River connects Lake Erie with Lake Ontario, as the waters of this vast lake system make their way to the Atlantic. Constantly eroding backward, upstream toward Lake Erie, Niagara Falls is itself only a temporary marvel: When the lip of the falls eventually reaches Lake Erie (estimated to happen a scant 30 thousand years from now, assuming a rate of water flow, and thus erosion, comparable to the present), Lake Erie will drain even faster, soon ceasing to exist, replaced by a raging river. All lakes eventually die as all flow out, and the edges of their containing basins are simply worn away.

Changeable though all terrestrial waterways may be, rainfall and the presence of open water dictates much of the basic character of local ecosystems. Wetlands—defined as terrestrial habitat that has standing water at least part of the year—are critical to the migration of waterfowl, shorebirds and other migrants. Particularly susceptible to reclamation (the draining of temporary wetlands and even of swamps for farmland, housing developments, and malls is currently a hot topic in the halls of the United States Congress), wetlands are a vital segment of the overall hydrological system of a given area.

Fish, fresh water mussels and snails, crustaceans, and insects and their larvae (many free-flying insects, like mosquitoes, black flies, and dragonflies, lay their eggs in lakes and streams, and the larvae live in the water until they metamorphose into adults) are the dominant freshwater fauna. Their health mirrors the physical state of the system: Long before a river becomes so polluted that its surface actually catches fire (as the Cuyahoga

River near Cleveland, Ohio did in 1969), the state of its fish and inverte-brates tells the story of increasing stress and of water progressively unsafe for human use.

Back in the 1970s, as a young professional interested especially in ex-tinct trilobites and their modern collateral kin, horseshoe crabs, I was dis-tressed, though not particularly surprised, to read that the Raritan River, which traverses New Jersey, reaching the Atlantic just south of New York City, had become so polluted that all but a handful of its animal species had disappeared near its mouth. There, the Raritan is an estuary (meaning that it has saltwater running in from the sea at high tide, mixing with the freshwater flowing outward from the interior). Virtually the only species left was the tough old horseshoe crab—a nearshore marine species well adapted to the less salty waters of estuaries but also a hardy beast, a true ecological generalist, capable of withstanding great swings, not only in salinity, but also in temperature, food supply, and now, apparently, in chemical pollutants that drove out nearly all the other species of its ecosys-tem. It is fortunate indeed that the hydrological cycle keeps going, that the system will cleanse itself once the pollutants are curbed. The lakes, rivers, and streams of North America on the whole are far healthier than they were a quarter of a century ago, and it is to be profoundly hoped that stan-dards will not once again be lowered in the name of economic freedom—a false economy indeed if our lakes and rivers once again begin to die.

Where wetlands meet the sea—as they do in all marshlands ringing a continent, or where the Everglades, a moving sheet of water through an open grassland, gives way to mangroves and then suddenly to ocean wa-ter—lies yet another critical aquatic region. Marginal marine marshlands are a nursery, both for open marine species that spawn among the reeds and roots, and for others that serve as their prey. With world fisheries in sharp decline, the persistence of these nurseries in a healthy state is crucial to turning around the progressive loss of many commercially exploited marine fish and invertebrates.

Threatened though they are by overfishing and pollution, the open oceans remain the last ecological bastion of relative normalcy on the planet. Nevertheless nearshore environments are especially vulnerable. Be-cause of human flotsam and jetsam (including medical wastes, virtually ru-ining the New Jersey shore season a few years back) or, particularly in the southern hemisphere, the dangers of exposure to ultraviolet radiation, beach closings are becoming commonplace.

As might be imagined, our grasp of true diversity beneath the waves is even murkier than the picture we have been able to put together for dry land. In the easily sampled intertidal and shallow-water continental shelf habitats, diversity in the oceans—at comparable depths—does seem to increase, as it does on land, as one moves from higher latitudes toward the equator. Likewise, diversity is commonly said to increase as one leaves the shoreline (where the ecosystems comprise largely ecological generalists, such as we find in the intertidal zone), reaching a peak in the more quiet bottom communities well below the wave base but still (at a depth of, say, 30 to 60 meters) well within the *photic zone*, where sunlight can still penetrate all the way to the bottom. Clams, snails, squid, crustaceans, sea urchins, and starfish head the list of the most conspicuous invertebrate elements, supplementing the local roster of sharks and fish in such communities.

On the ocean's abyssal plain, at an average depth of some 4,000 meters, diversity has always been reckoned to drop. Not much can live at such depths, withstanding almost unimaginable pressure in such unmitigatedly cold, dark bottom waters. To be sure, lantern fishes, giant squid, tile fish, and other strange denizens of the deep have long been known, and there are plenty of pictures of whole armies of ophioroids (skinny-armed starfish relatives) completely covering a sector of abyssal sea bottom. The basic picture of the ecosystems of the very deep oceanic plains has been one of relative modesty when compared with shallower regimes.

Thus, it has come as a bit of a surprise—at least to me—to read of recent expeditions utilizing deep-sea submersibles, recording lots of different species at various randomly selected sites in the Atlantic, with little or no overlap in species between regions separated by a few hundred kilometers. The old picture I have always had in mind was one of monotony and uniformity of the deep-sea bottoms, meaning that most species would be expected to be found virtually worldwide, with little in the way of geographic barriers to keep them apart or to encourage the emergence of new species. The deep sea was the last place I had expected to find relatively localized areas of endemism!

Based on discoveries beginning in the 1970s, perhaps we should have predicted that diversity would be higher and more localized than previously thought. At deep sea vents scientists found what must rank as the world's most unusual ecosystems, the only ecosystems fueled, not by the sun, but by radioactive decay emanating from the very bowels of Earth. Think of it: All organisms, save some crucial bacteria, rely on sunlight: they either photosynthesize their own sugars or gain their energy secondhand by consuming organisms that photosynthesize. Not so in the deep sea, where hot plumes of water belch out through cracks in trenches that extend far below the abyssal plains themselves. Chemoautotrophic bacteria, which are able to trap ambient heat through a sulfur-based metabolic pathway, have evidently replaced photosynthesizers at the base of this bizarre food chain, thousands of feet deeper than sunlight can penetrate. These bacteria in turn are eaten by giant filter-feeding clams and huge (up to 6 meters long) sedentary polychaetes—giant tube worms that filter food particles with an array of feathery tentacles.

These deep-sea vent faunas may be more widespread than previously imagined, and they may serve as nuclei for local hot spots of diversity all over the world's oceanic floors. Just as new phyla continue to emerge with some regularity (as the Cyclociliophorans did in 1995), so, too, do new ecosystems. The oceans—especially the deep sea—remain the least explored and least understood part of our planet.

CORAL REEFS Coral reefs are the marine equivalents of tropical rain forests. They are themselves quintessentially tropical. Reef-building corals are colonial organisms, able to secrete massive amounts of the hard skeletal mineral calcium carbonate. In this task they are aided by photosynthetic algae that live symbiotically in their tissues, providing the corals with the extra amount of energy needed for the arduous task of reef building. Such *hermatypic* corals live only on the equatorside of 40° north and south latitudes, wholly within the Tropics.

Reefs are organic structures full of nooks and crannies and make great hiding places for many small invertebrate and fish species. Thus, reefs are like nurseries of biodiversity, and literally hundreds of species might be represented at a single reef in the south Pacific. Some of the most colorful fish in the world are tropical coral-reef denizens. Reefs play important roles in the economic lives of local peoples, who gather sea urchins, clams, snails, and fish off the reef edge. Reefs play other roles as well, such as protecting shorelines, as we noted briefly in the preceding chapter; or, in a more negative light, posing hazards for marine navigation. (It was tricky maneuvering our 70-meter-long ship through the fringing reefs to get to Komodo Island in Indonesia to see the habitat of the fearsome Komodo dragon; the captain took it *real* slow).

The "miner's canary" side of reefs prompts my attention at this juncture. Compared to tropical rain forests, reefs may not be as extensive, or harbor such amazing diversity, or play critical roles in local and worldwide economies. But they are, in their own right, dazzling displays of high marine biodiversity, as anyone who has snorkeled over one, or watched any television footage about them, is well aware. Like tropical rain forests, they are fragile areas of high endemism. The recent, and still somewhat mysterious "bleaching" of corals (the deaths of the often colorful coral individuals, leaving only their stony white skeletons behind) is real cause for alarm. Local areas suffer most when reefs take ill, die, and soon crumble away. Corals clearly function as an early warning system, and the message from recent deaths not caused by silt runoff from overlogging, or oil spills is alarming. We might be seeing the early signs that the oceans, until now able to remain relatively pure by reason simply of their vast bulk, and the fact that humans live on dry land, are now feeling the effects of our presence on Earth. Just as the evident worldwide decline of most frog populations may be an early warning of an increase in ultraviolet radiation (or perhaps some other equally insalubrious atmospheric effect, see chapter 5), so too does the rise in otherwise inexplicable coral-reef die-offs portend ill for the oceans.

The oceans are absolutely vital to Earth. If any single system "runs the planet," it is the oceans, which basically determine weather patterns through interaction with the atmosphere. Oceanic plankton produce most of the daily supply of newly released oxygen, a supply that, if seriously impaired, would eventually (in about 12,000 years, according to some calculations) result in the reduction of atmospheric concentrations of oxygen. The reefs seem to be telling us that this system is beginning to run into trouble.

ALTITUDE MIRRORS LATITUDE: FROM QUITO TO THE GALAPAGOS
The general rule of thumb that says that biodiversity increases as one runs from the poles to the equator is closely mirrored as one runs down from the top of the highest mountain peaks to their very base. The phenomenon is most striking in the Tropics, where the *altitudinal* gradient in biodiversity neatly confirms what the *latitudinal* diversity gradient seems to be saying: Temperature and its fluctuations play the most important role in determining the distribution of biological diversity over the face of Earth.

Ecuador sits astride the equator—hence its name. The region around Quito, the capital, lies in the long valley between the east and west ranges of the Andes, at an invigorating altitude of some 2,750 meters. Though much of the region is under cultivation, and the seemingly ubiquitous eucalyptus trees have replaced native vegetation in many places, stands of native mixed forest are still within striking distance of downtown Quito. (I have seen eucalyptus virtually everywhere in the world: California, southern Africa, Madagascar, Ecuador, Indonesia—everywhere save its native continent of Australia, which I have, as yet, not had the good fortune to visit. Chosen for its fast-growing, thus commercial as well as ornamental properties, eucalyptus has had almost as deadly effect on local ecosystems, as other famous introduced species—such as the brown tree snake which wreaked such havoc after reaching Guam [see chapter 5]. Outside Australia, birds and insects for the most part cannot utilize eucalyptus for food and shelter as readily as they can the native plant species to which they are adapted (not only animals pose ecological hazards as invading species).

Several magnificent snow-clad volcanic peaks are visible from downtown Quito. They have the classic Mt. Fuji conical look typical of continental volcanoes. One, Volcan Cayambe, literally sits on the equator. (Standing only 6,700 meters above sea level, it is, nevertheless, the tallest peak in the world, taller than Mt. Everest at 8,848 meters above sea level, *if tallest means farthest from the center of Earth*. Earth bulges, making the distance from sea level to the center of Earth some 13,000 meters farther at the equator than at the geographic poles).

I visited another volcano, Volcan Cotapaxi, just south of Quito. After a scenic journey along the intermontane plain, we began to climb up to 3,500 meters, eventually coming to a dense forest of conifers. The forest began at the entrance to a National Park and held a protected herd of increasingly rare llamas. Emerging at the other side, we found ourselves, nearing 3,700 meters, in a region of broken forest and low shrubs. A little farther and we were above timberline, in grassland dotted with gorgeous wildflowers. Above us loomed the peak itself, snow-clad like its fellow peaks in the chain, the snowfields beginning at some 4,300 meters and rising all the way up to the peak. At the little lake beside the road, by now over 3,700 meters up, we saw Andean lapwings and speckled teal—adapted to the frigid, high-altitude lakes and ponds.

We were clearly on a high-altitude version of the tundra. Above tree line the world over, "alpine" meadows succeed predominantly coniferous forests, mirroring exactly the sequence of vegetational realms we have already encountered moving along a latitudinal transect. Tree line might be relatively higher up the mountains along the equator (Mt. Kenya and Mt. Kilimanjaro are exact African analogues to our South American examples) than in the Himalayas and other higher-latitude mountain systems, but the principle is the same everywhere: Changes in altitude produce ecological differences very similar to those seen with changes in latitude. Small wonder the two words—at least in English—are barely-different anagrams of one another!

As might be expected, high-altitude species tend to be ecological generalists like their high-latitude counterparts, though with high altitude comes the additional problem of oxygen availability. Andean peoples, for thousands of years living well above 2,100 meters (the elevation of the famed *ruins* of Machu Pichu), have in a relatively short time span become adapted—in the literal, genetic, and evolutionary sense—to the thin air of these lofty domains, while casual visits by biologists and tourists require a much more rapid physiological adaptation that, in some cases, never quite fully happens. In the long run, species become adjusted, and thereafter none suffer from any long-term symptoms of altitude sickness.

One factor definitely separates mountain-top living from life in the forests and tundras of the higher latitudes: On the continents, as we have seen, species—such as wolves and moose of the northern hemisphere—tend to occur in large numbers of individuals, and occupy large regions of a continent, often actually living completely around the globe. This is not true in montane areas, where tundra and even high coniferous forests are separated from one another like islands in the sea. The greater the distance, the more difference in the flora and fauna between any two mountain tops, mirroring in this instance, not the higher-latitude continental habitats, but rather oceanic islands. Thus, mountains—like islands such as Madagascar—are likely to be rich in endemic species. Endemics, with their relatively narrow distribution, are far more vulnerable to extinction than far-flung species, no matter how ecologically generalized they might otherwise be.

I left the Quito region reluctantly, but was eager nonetheless to get to the next destination: the Galapagos Islands, the mecca of all evolutionary biologists, known as a virtual laboratory of evolution, islands with their own share of endemics. These islands, perhaps the most sacred symbol of living diversity on Earth, are also very much threatened by sheep let loose by sailors to serve as a living larder, and by the expansion of the usual roster of modern human activities.

En route, I saw little from the air—cloud cover obscured the land, the waves obscured the marine communities—but had I had the opportunity, I would have seen the resumption of tropical forest down the western Andean slopes, extending toward the Pacific coast. I might have glimpsed some grasslands, and later the deserts that fringe so much of the Pacific coastal lowlands. The same rules apply: The lower the land lies, the more it resembles lower-latitude ecosystems.

On a submersible, or at least on a ship equipped with radar and other devices to probe the depths, I would have seen intertidal communities quickly giving way to shallow-water benthic (bottom-dwelling) communities of clams, snails, octopuses, sea urchins, crabs, and many other invertebrates. Then, a quick drop-off down the continental slope takes us to the abyss and eventually to the Galapagos rift, with its own vent fauna, supported by bacteria.

The volcanic islands of the Galapagos, massive shield volcanoes with a much more gently arced shape than the conical volcanic peaks of the Andes (still only 1,400 kilometers directly to the east as the "crow" flies—there being no crows in the Galapagos) rise out of the water, and we are now back to terrestrial ecosystems. The Galapagos lie on a *hot spot*, a plume of active upwelling of oceanic crust lying at the juncture of three separate crustal plates. The oldest islands lie in the southeastern reaches of the chain, the youngest in the northwest. As new islands are created, they are populated by species that arrive there usually completely by accident, carried by winds (certain types of seeds, insects, or even birds); tides (marine creatures, or even terrestrial species caught on floating logs); or the courtesy of other species (e.g., seeds in the guts of birds flying over the new space).

There, natural selection may well modify these immigrants, isolated from their parental populations, provided they survive and provided there is sufficient genetic variation in place in the pioneer population. They will evolve to fit the slightly different environment of the new island, often over surprisingly brief periods of time, modifying their features and their reproductive adaptations, in short, becoming new species. The patterns of differentiation of related species are especially clear in the Galapagos: Darwin, for example, noted that the land tortoises seemed to be different on each of the major islands of the chain.

Nonetheless, the islands are potential hot beds of extinction as well. The Galapagos are a microcosm of Earth itself: A generator of living diversity, an "entangled bank" of existing diversity, and a fragile series of ecosystems, staffed in no small part by endemic species, facing imminent collapse. Should extinction claim many of the Galapagos species, it is to me small comfort to realize that, thousands of years in the future—but only after humans leave the Galapagos alone—evolution in a sense will "fix" things: Some species will hang on and bounce back, and from them, other new species will evolve. The point is that, however far removed from local ecosystems we have become, we are very much a major part of the global ecosystem, and thus the fates of all the local ecosystems that add up to the global system very much matter to us. We must do all we can to stem the rising tide of the Sixth Extinction—whether in the Galapagos, symbolic capital of biodiversity, or for that matter anywhere and everywhere else in the world.

ECOSYSTEM DIVERSITY AND THE DYNAMICS OF CHANGE

Botswana's Okavango Delta is changing. Drought is reconfiguring the landscape, and water no longer flows as readily and regularly in the Boteti River, the once dependable overflow outlet of the great swamps.

The problem is not simply a lack of rainfall within northern Botswana itself, nor, for that matter, human diversion of the Okavango River, which feeds the swamps from the north—at least not as yet. Rather, rains have diminished at the source, in the Angolan highlands. The larger islands of the delta, already open savannas in their centers, are beginning to coalesce into still larger expanses of open grassland. It is confusing—and disconcerting—to fly over the delta and see arcs of riverine forest standing out far from water, like British hedgerows separating adjacent open farm fields, but that's what happens when the floods fail, when waters are simply too skimpy to spill over what, until recently, has been an annual floodplain. The southern Okavango Delta is becoming more extensive open savanna.

Nature abhors a vacuum. When expanses of grasslands are not covered by annual floods, the grass species covering the region change, and as a nearly instantaneous ecological reaction, the mammals and birds change. I was very surprised to see, on a recent trip to the very heart of the deep-water permanent delta, several female cheetahs and young occupying the now-dry floodplain fields just off the river. It would not be remarkable if this were the first sightings of cheetahs in the delta by an infrequent visitor such as myself. It is far more arresting and significant to learn that these were the first cheetahs seen by many of the veteran guides and camp managers who have spent decades in the delta without so much as a glimpse of one.

Cheetahs are icons of the open African plains. A host of other more subtle changes have occurred: Spotted cats are on the rise in the same region; and long-tailed starlings have begun to outnumber their close relatives, Burchell's starlings, in many of these same central Okavango Delta habitats. If drought persists, if the floods continue *not* to come, these subtle changes will snowball into a profound change in the delta's ecosystems.

Are these changes good or bad? They are neither—they are just changes. Nothing lasts forever, and that is absolutely the case in terms of the physical climate, of temperatures and patterns of rainfall. True, wetlands and deep-water ecosystems typical of the Okavango—dwindling remnants of ecosystems once known in ancient Egypt and even further back in the remote mists of geologic time in Olduvai Gorge—are so scarce that their disappearance in Botswana *would* be lamentable.

But as we have already seen, when climates change, forcing ecosystems to change, what really happens is the expansion of one form of habitat at the expense of another. More likely than not, those "displaced" habitats themselves expand elsewhere. Like the species that staff them, ecosystems move around in response to climate change. Not just birds, such as tufted titmice, cardinals, and Carolina wrens, or mammals such as opossums, but also many plant species have been moving steadily northward as the climate of North America has been steadily warming over the past century. Just as we chart extinctions through counting individual species, so do we tabulate the movements of separate species: cheetahs and long-tail starlings moving into the Okavango Delta, tufted titmice and opossums living in New England. What is really happening is climate-induced transformation of ecosystems themselves.

If change is natural, then why decry human-induced transformation of the surface of Earth? The answer is simple: By transforming grasslands and forests into farmlands, cities, suburbs, and shopping mall complexes, we humans are not simply displacing ecosystems elsewhere, but rather actively destroying them, shrinking the habitat necessary to support a vast range of species. Agriculture, for example, has removed all but a fraction of the remaining giant panda habitat since the invention of writing, and hence historical mention of pandas.

If size of ecosystems is relevant for continued survival of constituent species, so is the related issue of *degradation.* A woodlot in suburbia may look like normal eastern North American woodland, but in reality be riddled with introduced (nonnative) plants and animal species, and otherwise contain just a fraction of the plant and animal species needed for normal, healthy ecosystem functioning. Ecologists have recently concluded that such degraded ecosystems, where diversity has been diminished through human interference, are less stable—less able to withstand additional stress —and thus prone to suffer rapid changes.

With fewer than normal original numbers of different species present, degraded ecosystems are also considered less productive, meaning that less energy is generated with a sustained flow through the system. Thus, degraded ecosystems are by definition less robust, less healthy than they were in their original condition.

These issues are complex. We must always bear in mind that, 10,000 years ago, there simply were no forests in what is now New York City because New York was still buried under an 800-meter thickness of glacial ice. Change is normal, and the "primordial" woodlands of northeastern North America are not all that old as things go in geologic time.

Nor should we conclude that moist tropical ecosystems are inherently better than desert or tundra systems. More solar energy supports productivity in the Tropics than in the higher latitudes, and there is no reason to suppose that tropical systems are more stable than tundra systems. Greater biodiversity within systems is "better" only when we compare a pristine system with its human-degraded counterpart.

Ecological change is natural. Ecosystems change and move around. Species do not become extinct as long as they can literally keep pace with the change, as when cheetahs take advantage of their new opportunities and move into the expanding grasslands of the Okavango Delta.

But change—if too rapid or absolute—can, and indeed does, lead to extinction. If the Okavango Delta dries up completely, I fear for the future of the slaty egret, a magnificent loner that spears frogs and fish in shallow floodplain waters nowhere else. Extinction is almost always a consequence of too much or too rapid ecological change.

Chapter 5 BIODIVERSITY —A THREATENED NATURAL TREA- SURE

The San people of the Kalahari have no trouble whatever understanding the value of biodiversity: Until fairly recently, the San had been living in small bands wholly within their local ecosystems. All their food, their clothing, their shelter, their medicines, their cosmetics, their playthings, their musical instruments, their hunting weapons, everything came from the productivity of their surroundings, the plants and animals on which they completely depended for a living. ❧ *Why, then, is it so difficult for most of us in the industrialized nations—urban dwellers but also rural farming folk—to grasp the significance of biodiversity? The answer, I* truly believe, is that we have simply forgotten what the San and all other hunter-gathering peoples still know. We have forgotten because of a profound and radical change in our relation to the natural world that came as a direct consequence of the invention of agriculture. We need to understand how people fit into the natural world—both as hunter-gatherers *and* as agriculturally based industrialized societies before we can assess realistically what biodiversity means to human life.

For the first time in the entire history of life, one species, *our* species, *Homo sapiens*, has stepped outside of the local ecosystem. Agriculture changes the entire relation between humans and everything else living in the vicinity. To plow a field, to cultivate a handful of crop species, means the destruction of the dozens of native plant species that would otherwise

[Figure 41] *The Florida Everglades. Once literally a moving sheet of water flowing through grasslands towards the sea, the Everglades became the object of massive "reclamation" efforts to divert water and create agriculturally useful land. This scene depicts such human modification alongside a patch of relatively pristine habitat—in which a great egret stands (once endangered, along with snowy egrets, through over-hunting for their feathers) while a wood stork flies overhead. Public outcry over the degraded condition of the Everglades has prompted a new initiative to restore, as fully as possible, the original Everglades wetlands system.*

be there naturally. No one ever heard of a "weed" until we began dictating what limited number of plant types we wanted to be growing on a given plot of land.

What is the difference between living off the natural fruits of the land and living off those we grow ourselves? The answer is simple: To live off our own cultivars, we must disassemble original ecosystems. There is very little native North American prairie left in the Midwest, and therein lies the gist of the dilemma. Most of us genuinely think we don't need prairie at all. We see prairie as simply underutilized terrain. We even tend to look at marshes like the Hackensack meadowlands and see instead Giants Stadium and the potential for still more entertainment and business complexes rather than a New Jersey tidal wetland full of cattails, migrating birds, and larval marine life vital to the restocking of the marine fisheries on which we still so heavily depend. [Figure 41] The Botswanan cattle industry looks at the grasslands of the Kalahari—and increasingly, the greener pastures of the Okavango Delta itself—as underutilized rangeland. We have come by this outlook honestly: Having stepped outside local ecosystems so successfully, starting 10,000 years ago, we have come to think that we no longer need prairies, wetlands, or any other kind of natural habitat.

Agriculture has been a stunningly successful ecological strategy. Though famine has stalked the enterprise from its inception (there really is no such thing as complete control over food supplies, or anything else for that matter), the best indicator of ecological success is growth in population numbers. Estimates vary, but it seems likely that there were *no more than 5 million* humans on the planet 10,000 years ago. We had recently completed our spread throughout the globe by then, but we were still organized into relatively small groups as hunter-gatherers, still utterly dependent on the productivity of the local ecosystems in which we all continued to live.

The upper limit on human population numbers back then was set by the same rules that govern the numbers of all other species: Each population is limited by the environmental carrying capacity, the number of individuals that, on average, a local habitat can support, taking into account available food and nutrient resources, and other important factors, such as prevalence of predators and disease-causing microbes, and even more general factors, such as climate and rainfall. Each local population is fixed at some fluctuating number, usually 30 or 40 individuals maximum, as was usually the case with San and other hunter-gathering human beings. Thus, the total number of individuals of any species is the average size of its local populations multiplied by the number of those existing populations.

Agriculture popped the lid off natural regulation of human population size. No longer limited by the inherent productivity of local ecosystems, human agricultural societies began to expand immediately. Agriculture enables a settled existence—and as populations began to grow, as patterns of political control and the division of labor began to emerge, human life rather quickly took on a semblance basically familiar to those of us living in even the most advanced of modern societies. [Figures 42A-C] Nor is this, of course, a "bad" thing: All of the great accomplishments of human civilization spring from our forsaking the local ecosystem and adopting agriculture as perhaps the pinnacle of our culture-dominated mode of making a living in the world.

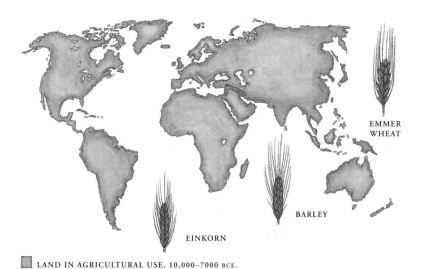

EMMER
WHEAT

BARLEY

EINKORN

▓ LAND IN AGRICULTURAL USE, 10,000–7000 BCE.

[Figure 42A] *Land under agriculture between 10,000 and 9,000 years ago. Agriculture was invented at a number of different times and in different places. Agriculture means taking control of your food supply—deliberately growing plants and tending domesticated animals—a perfectly rational, and indeed, triumphant human ecological invention that was to have many consequences.*

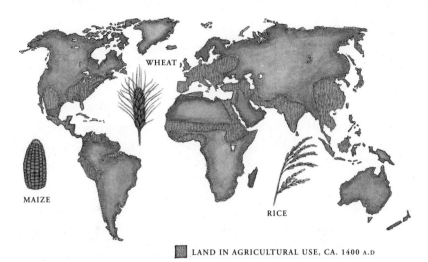

LAND IN AGRICULTURAL USE, CA. 1400 A.D

[Figure 42B] *By 1400 AD, agriculture had become far more widespread. One major consequence of the invention of agriculture was the explosion in the human population. Another was the rise of city-states—with division of labor (economic specialization) —including the invention of writing, the ever-accelerating growth of knowledge, and with it great creativity in the arts, science, and technology. Yet one other consequence: the noticeable conversion of natural terrestrial ecosystems for arable land.*

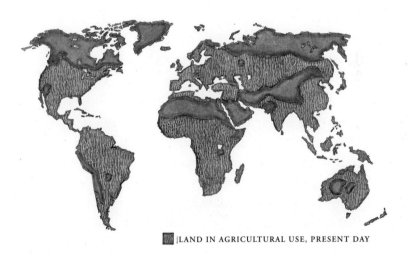

|LAND IN AGRICULTURAL USE, PRESENT DAY

[Figure 42C] *The present day. The trend has continued, and virtually all of the arable land has been converted to agricultural use. Indeed, traditionally non-agriculturally useful lands have also been converted—as wetlands are drained and deserts made to bloom. Not to mention the spread of cities, and especially mall-ridden suburbia (a phenomenon by no means restricted to the United States). I look at this map and wonder how the biota of 10,000 years ago has managed to survive at all!*

If high culture is one signal of our success, so too, in the time-honored measure of ecological success, is our geometric increase in the number of individual living humans at any given moment. If there were perhaps as many as 5 million people alive 10,000 years ago, there are now nearly *6 billion* of us. We are engaged in a perpetual race to feed ourselves, and every time we come up with a clever expansion of agricultural technology—whether it be crop rotation and efficient plowing techniques a few centuries ago, or biotechnological manipulation of the genetics of crop plants today—human population numbers expand right along, so that there are always people on the brink of starvation somewhere.

The sheer bulk of human numbers—this 6 billion and ever-expanding, probably nearly doubling to over 10 billion by mid-21st century—is wreaking havoc on Earth, on its species, ecosystems, soils, waters, and atmosphere. We are the current cause of this great environmental crisis, this threat to the global system that looms even as we approach the Second Millennium. We have created the biodiversity crisis, the next great wave of mass extinction that promises to rival the five greatest extinctions of the geologic past—The Sixth Extinction.

Our forsaking local ecosystems is only half of the current human ecological equation. There is something else that is absolutely unique to the human ecological condition, something never before developed by any species in the entire 3.5-billion-year history of life on Earth: *We are an internally integrated global species.*

Other species, to be sure, have also spread around the globe. Many, like the Norway rat, are commensals on human settlements, meaning that they have adopted a life dependent on humans, and have simply accompanied us as we have spread over Earth. All these other species remain dependent on the largesse of their local surroundings—in the case of rats, in the contrived "ecosystems" of human habitations. The rats that raid my backyard bird feeding stations care not and know not of their fellow rats in their native Europe. Their economic lives—the business of developing, growing, making a living, and staving off death—and those of all other species are played out in local, circumscribed areas, the confines of the local ecosystem. What little connection exists between far-flung rat populations is genetic, as occasional contacts allow genes to be exchanged. Because Norway rats still hitch rides on ships, we can confidently assume that Norway rat genes still flow back and forth across the Atlantic ocean. But that is the only link: Those rats eating bird seed in New Jersey have no economic connections with the rats pilfering orange rinds from an Oslo garbage can.

Our species is different—and crucially so. Each of us is connected, in a direct economic sense, with all other human populations on the globe. We import, we export. Indeed, as we shall see, much of the direct, utilitarian value of life's exuberant diversity lies in our economic connectedness with our fellow humans in disparate corners of the globe. People have been exploiting far-flung biological resources long before Hannibal used elephants to cross the Alps in 218 B.C., and Marco Polo opened up the spice trade between western Europe and the Orient 700 years ago. We now exchange an astonishing *1 trillion* dollars worth of goods and services *daily* among ourselves. We are an amazingly inner-connected species.

This human global economic interconnectedness has once again altered our stance vis-à-vis the natural world. Since we took our first agriculturally induced steps away from local ecosystems, we have been in transition. With plenty of space in which to expand, and with relatively low initial population numbers, our leaving local ecosystems really was tantamount to declaring independence from nature itself. But our very success, especially as measured in our virtually out-of-control population growth, contains a real joker in the deck. We have become an integrated species, and, *for the very first time* in the history of life, a species that exerts a coherent economic role in the natural world.

How does this 1 trillion dollar per day international exchange in goods and services impact the natural world? For a one-way example, just think of U.S. imports of oil from Saudi Arabia, beef from Argentina, wine from Chile, and camcorders from Japan. Our collective enterprise exerts an influence on the globe as a whole. The average citizen of developed nations, such as the United States, consumes some 30 times as much in goods and services as a citizen of an underdeveloped, Third World nation such as Bangladesh. Buying veneers from the Tropics supports the destruction of rain forests, and the disparity of monetary resources—economic clout— between rich and poor nations masks in part the true significance of the impact of a nation's total population on the global system. In other words, we must multiply the U.S. population of 280 million by 30 to get an accurate grasp of how we compare with Bangladesh's 70 million people in terms of our economic and ecological impact.

There truly is a global system. James Lovelock called the system "Gaia" (Greek for "Earth"), and maintained that Earth in its entirety—its atmosphere, its soils and rocks, its waters, and the rind of life clinging to its surface—is itself like a gigantic living thing. Although many biologists rightly object to the overly fanciful simile of Earth as a living being and prefer the word Biosphere, there is something very important about the basic Gaia message. The world's local ecosystems are simply portions of Earth's surface that house collections of different species of microbes, fungi, plants, and animals, collections that differ from place to place because of different physical characteristics of the environment.

The distinctions between local ecosystems can be sharp and dramatic—as when I once placed one foot in the rich alluvial soil of the Nile Valley floodplain (now largely under cultivation in any case) and the other on the dry, rocky, and contrastingly quite barren Sahara Desert. In many other situations, ecosystems grade almost imperceptibly into one another, as when a mixed hardwood woodland grades up a mountain slope into a stand of predominantly conifer trees. It is often difficult to draw a hard and fast line between local ecosystems.

That is the point: The flow of energy between species of microbes, fungi, plants, and animals in any local setting is connected *laterally* with adjacent regions. Local systems are linked into regional systems. The threatened pine belt of the southeastern United States is a prime example: It was first exploited for the production of naval stores (for trading purposes as well as domestic consumption), and is now attacked by the inevitable expansion of population with its concomitant malls, golf courses, and housing developments. Regional systems are themselves linked up into continental systems. And the whole is integrated through patterns of atmospheric and oceanic circulation. There is a global interconnectedness to the world's environment, and Earth's living creatures are very much a part of that interconnectedness.

There is a global environmental system—the Biosphere—which is roughly 4 billion years old. And now there is a global species, *Homo sapiens*, which is playing a coherent role within that global system. This is something very new in the history of life. Indeed, it is very new even in the history of *Homo sapiens*. That is why, in forsaking local ecosystems, we have not managed to escape the clutches of the natural world after all. We have to realize that, in the past 10,000 years, we have redefined the global system as our own *mega-ecosystem*.

The reason it is so hard for many of us to see the values of biodiversity, then, is simple: We stopped living in the confines of local ecosystems which naturally made us think we were no longer part of the natural world. What really happened over these last 10,000 years was not so much a leaving of, but rather a redefinition of who we are and how we fit into the rest of the world in an ecological sense. We are still very much a part of the natural world despite the entirely new relationship we have established with it. This relationship is very much a two-way street. As we have already described, it is the impact of humanity, with our huge and ever-expanding population, that underlies the current building wave of extinction that threatens to envelope Earth's ecosystems and species. That is an aspect of *our* impact on the global system. It is fairly easy to see, though the consequences of this impact are as yet not fully realized.

What is far less well understood is the other side of the coin: the impact of the *global system* on us. Because we are still stuck with the notion that we have escaped the natural world, few of us see the dependence that our species truly has on the health of the global system. The main reason we should fear the Sixth Extinction, I truly believe, is that we ourselves stand a good chance of becoming one of its victims. If it might be difficult literally to extinguish, say, 10 billion people, I need only point to the distinction between cultural and actual biological extinction. We might well avoid literal biological extinction—but our cultural diversity, and, for the developed nations, our high standards of living, are very much at risk.

Thus, the dual theme of this narrative: We humans are having an increasingly devastating impact on the biological and environmental systems of the entire globe, and yet we rely strongly on the integrity of the global system for our continued existence. We need to understand how the global system affects us as much as we need to see how we affect that system. Only then will we come to see the very real threat that the Sixth Extinction poses to ourselves. We need to strike a balance between ourselves and the rest of the natural world.

THE VALUES OF BIODIVERSITY Three

themes crop up in everybody's lists of why diversity matters. We have already encountered all three in passing. They are (1) utilitarian values (such as medicine and agriculture); (2) ecosystem services (vital functions such as the continued production of atmospheric oxygen); and (3) moral, ethical, and esthetic values.

Just as most of us don't know how our telephones, TV sets and computers work, we really have only the vaguest idea of where our foods and medicines come from. Harder yet to understand is the significance for our very existence of species and ecosystems which seem to just sit there and provide no obvious product for us to eat, use as fuel, or stock our medicine chests. Vaguer still is the calm sense of joy and simple belonging most urbanites experience with a simple walk in a woodlot, through a meadow, or along a clean shoreline. Yet these three categories of the effects of the living world on human life are absolutely crucial to modern and future human life on planet Earth.

UTILITARIAN VALUES OF BIODIVERSITY When asked how

many species humans routinely utilize in their daily life, most people (including most professional biologists) say, at most, perhaps one or two hundred. The correct answer is at least 40,000: Globally, each day we depend on over 40,000 species of plants, animals, fungi, and microbes. I am counting here only those species that we are deliberately exploiting. Still others, such as the microbe *Escherichia coli*, which lives by the millions in our intestines and is absolutely vital for normal digestion, are, fortunately, simply there.

Many of us think that food comes from the grocery store, and have little idea of its ultimate provenance. If some of us realize that spaghetti comes, not from trees but from wheat flour, we still tend to think that the Agricultural Revolution is long since complete, that we have already abstracted from nature all the plant and animal species that we are ever going to farm. We think that whatever improvements in crop yield and disease resistance—two critically important factors in the ongoing race to feed the 250,000 extra mouths we are currently adding each day—can come strictly from improved breeding techniques, and especially from the seeming magic wrought by the recently developed techniques of biotechnology.

Nothing could be farther from the truth. Here, a direct analogy with the natural world is apt: Evolution works through natural selection, the process Darwin (and Alfred Russell Wallace) discovered. On average, the organisms that thrive best will survive and reproduce, passing to their offspring the very traits that allowed them to flourish. Breeders do the same thing, allowing only those sheep, say, that have the woolliest coats to reproduce in the hopes of producing future sheep with even thicker coats than their forerunners had.

But selection alone—whether natural or artificial—will not do the trick. Another ingredient is required: the presence of genetic variation. You can only select from an assortment of different traits. Once you have gone as far as you can in selecting from the available range of genetic traits, the process, inevitably, comes to a halt. The reason why evolution did not stop billions of years ago is that spontaneous genetic changes—mutations—occur each generation, renewing and increasing genetic variation.

Biotechnology allows us to inject genes directly into domesticated plants and animals. At first glance, it seems that we have co-opted nature, once again substituting a clever bit of technology over a chancier and slower natural process. But the genes we insert to produce, say, frost-resistant strawberries, have to come from somewhere. You can't just go to a molecular biology facility and ask them to invent a gene that will make strawberry plants hardier. No one has the faintest idea what that gene would be, what its precise instructional coding would be, or where it might be inserted into the chromosomes of the strawberry cells.

Biotechnology works the old-fashioned way: One must first find a genetic feature that performs the desired function, before it can be extracted, manipulated, and inserted with the marvels of modern biotechnological technique into the stock where you would like to see that desired effect expressed. That means we must find genetic variation in the usual place: in nature, in wild versions of domesticated species, and in their nearest relatives. For many crop plants, there is an additional ace in the hole: The centuries, indeed the millennia, that farmers have been patiently tilling the land, sowing seeds, and harvesting crops that are bountiful one year, skimpy the next, have seen the emergence of countless *landraces*, local varieties of corn, wheat, tomatoes, etc. that seem to do best in a particular combination of local soil and climate. The history of agriculture has itself produced, through simple artificial selection, a vast storehouse of genetic variation.

All that variation is under serious threat. As science reporter Paul Raeburn recounts in his book *The Last Harvest* (1995), destruction of ecosystems in the wild threatens to obliterate countless species that are close kin to vital agricultural crops. He tells of the combination of skill, persistence, and luck that has enabled botanists from the United States and Mexico to locate a previously unknown species of wild corn, *Zea diploperennis*. This rare and rather unprepossessing plant promises to enable agricultural geneticists to abstract its genes, which convey resistance to a wide assortment of corn diseases.

The alarming coda to an otherwise encouraging story of the importance of natural genetic variation in wild species to our collective agricultural effort is that *Zea diploperennis* almost certainly would have become extinct within at most a few decades as its limited natural habitat in Mexico's Sierra de Monantlán was suffering precisely the same sort of conversion (for agricultural use!) that we are witnessing around the entire globe. How many other wild relatives of domesticated species have we already lost, and what will the effects of that loss be as we struggle to feed increasing billions of people over the next several decades?

Raeburn notes that a similar fate is meeting thousands of landraces. We lose species in the wild as we convert land for agricultural and other uses. We are losing landraces for a different reason: The switch from small single-family farming to large-scale agribusiness, coupled with recent dramatic advances in biotechnology, means we have begun to plant only a few "super" varieties of crop plants, apparently for good reason, as the crop yields have risen, and the quality remains high all the way to the dining room table.

There is a downside to all this success. Big agribusiness has seen huge increase in both fertilizers and pesticides, with their ongoing deleterious effects on soils, rivers, and adjoining ecosystems. The loss of hard-won genetic diversity of these landraces also poses a deep threat to continued agricultural success in the future. It just doesn't do to put all your genetic eggs in a single basket—allowing varieties with all sorts of as yet-unexploited valuable features to disappear in the rush to concentrate on a few, genetically homogeneous strains.

Genetic diversity is the key to past, present, and assuredly future agricultural success. It is the key, as well, to our utilization of virtually all natural products. The medicines in our pharmacopoeia are as compelling an example as the agriculture story. Although we might be aware, in a vague way, that aspirin was originally extracted from the bark of willow trees, and that Europeans first learned of the drug through contact with native Americans, few of us have any idea of the extent to which indigenous healing practices, and the most sophisticated research and development efforts of the world's biggest pharmaceutical companies, rely on the genetic diversity of wild species.

Graphic examples of the importance of wild plants to the development of new drugs have recently become famous: The Madagascar periwinkle, a wildflower and close relative of a common decorative horticultural variety, has yielded a drug that has proven effective against two forms of childhood leukemia. Taxol, extracted from the bark of the Pacific yew, is now an important part of the chemical arsenal marshalled against ovarian cancer. We have already encountered the drug extracted from the seed pods of the sausage tree in the Okavango Delta and other African locales, a drug recently shown to be an effective agent against skin cancer. True, taxol and other potent drugs can be made synthetically in the laboratory. The point is, though, that we first have to know about the existence of these compounds before we can make them. It would be folly—and horrendously expensive and time-consuming—to sit around a laboratory randomly cooking up compounds in the hope that one of them might prove useful to combat a particular disease.

As you might expect, local peoples who are closely tied to the land, and have not forsaken the old hunter-gathering mode for agriculture, have a vast storehouse of knowledge about the natural world around them. Scientific explorers have been repeatedly struck by the detailed knowledge of local peoples concerning the plant and animal species in their environment. For example, a new species, the golden bamboo lemur (lemurs are primitive primate relatives of monkeys and apes), was discovered by Western-world primatologists living in a section of western rain forest in Madagascar in the 1980s. The local peoples had known of its existence and the fact that it was different from its close relative, the greater bamboo lemur, which the Western biologists had confused it with. They could tell just by listening to its nighttime cries coming from the forest that the golden lemur is a distinct thing—what we westerners call a distinct "species."

More graphic still are the accounts from earlier expeditions. While collecting birds in the 1930s in New Guinea, the famous ornithologist Ernst Mayr found that the local tribesmen knew all the species that he could locate and actually could point to two confusingly similar species and tell him they were different. Nowadays, ethnobotany and ethnozoology are important areas of research—not least for what they reveal of the utility of plant and animal species already well-known to indigenous people. Local expertise about native plants and animals has other implications, as well. When we compare lists of plants and animals drawn up by local peoples with those of professional biologists, it confirms our notion that species are real entities in the natural world, not just figments of Western-world classificatory imaginations. Local expertise can also dovetail conservation efforts with the economic needs of indigenous peoples: for example, by paying locals to act as guides in conservation reserves, or to serve as parataxonomists helping in the sorting and identification of species—the "elemental particles" of biodiversity—in biologically poorly known regions.

The world holds far more than the 40,000 or so species currently being utilized on a daily basis. That is why the exploratory research efforts of the chemical and pharmaceutical industries must go beyond simply cataloguing the experiences of local peoples. Although we have no precise idea of how many plant, animal, fungal, and microbial species populate the planet, there are at least 10 million of them. The living world is a vast cauldron of genetic variation: Most of it remains entirely unknown to us, yet much is undoubtedly of great potential use.

For good reason, much of the exploratory research has been focused on the tropical rain forests. Most of the terrestrial species of our planet reside in the Tropics, and tropical forests are disappearing at a frightening clip. Estimates vary, but 30 hectares per minute now seems, if anything, to be an underestimate. More recently, however, some attention has been shifted to the sea, the last great earthly frontier. We are, of course, ourselves a terrestrial species, having abandoned the sea to take up life on land some 350 million years ago. Until recently, our direct utilization of sea life has been restricted to fishing and to hunting marine mammals. This last great vestige of a hunting-gathering mode of existence until recently threatened to extirpate many whale and seal species and, as we have already seen, now threatens to collapse the most productive fisheries in the world.

Corals and sponges are but two of the major groups of marine invertebrate animals that live firmly rooted to the sea floor. They don't move around, so they can't escape when a predatory fish or crab comes by and tries to bite off a piece. These sessile creatures have evolved a stunning array of chemical defenses against such attacks—defenses that have recently begun to attract a lot if attention from the chemical and pharmaceutical industries.

The case for the great diversity of living species as a storehouse of vital genetic variation is crystal clear. We have relied upon that variation increasingly since we developed agriculture, even as it has indeed seemed that we were abandoning nature. That reliance on the natural genetic storehouse will only increase as time goes on, a compelling reason why we must arrest the destruction of ecosystems and species that right now is systematically dismantling and destroying this vital resource.

Specific utilitarian uses are only part of the story. Ultimately what might prove even more crucial is the simple overall health of the global system: the purity of the air, the balance among carbon dioxide, oxygen, nitrogen, and other gases of the atmosphere; the quality and circulation of water; the vital cycles of carbon, nitrogen, phosphorous, and other atomic constituents of our bodies. In short, in fouling our nests, in destroying ecosystems, and driving many species to extinction, we are beginning to approach a limit on how much of the global living system—and we ourselves—can actually survive. In the long run, the most valuable aspect of diversity may well be the ability of our species to continue to live on the planet.

EARTH'S ECOSYSTEM SERVICES Why, one might reasonably ask, need we worry about the health of local ecosystems if we ourselves in large measure no longer live within them? Why can't we continue our 10,000 year course of habitat conversion and ecosystem destruction now that most of us no longer look to local renewable resources in our daily lives? Can we not live in a world wholly of our own cultural devising—

without all but a few of the world's species, on which we realize we have come to depend?

No, we cannot. We have emerged at the other end of the 10,000-year honeymoon with agriculture—and the consequent explosion in our population numbers—and have begun to see that we are part of the global system, after all. Earth comprises a global system of interacting elements: the atmosphere, the lithosphere (soils and rocks), the hydrosphere (oceans, lakes, streams), and the Biosphere—all of life. That global system is the summation of all those local systems interlocked across the entire face of Earth. Earth is our home—where we were born, where we live now and—space-travel fantasies notwithstanding—where we will have to stay if we have any chance of long-term survival.

What effect does the Biosphere have on us? What does the Biosphere do for us? Simple, essential and downright fundamental things—things that we mostly don't see, appreciate, or fully understand—without which, life on Earth for all species, including ourselves, would be completely impossible.

Take the air we breathe. The atmosphere close to Earth's surface is mostly inert nitrogen (79%), which in itself is a good thing, as an atmosphere richer in oxygen than it already is (20.9%) would literally fan the flames of out-of-control wildfires. When we talk about the air we breathe, most of us mean oxygen. Oxygen is absolutely essential to all but a very few forms of microbial life. Some bacterial species use alternative chemical pathways to break down the nutrients on which they live, but all the rest—most microbes, plants, fungi, and animals, including human beings—require a constant supply of oxygen just to exist.

Where does atmospheric oxygen come from? [Figure 43] With billions of organisms taking in oxygen and expelling carbon dioxide, surely we would soon deplete this essential resource. The answer, of course, is photosynthesis, the process whereby some bacterial and other, more complex microbes, as well as all green plants, trap solar energy by producing sugars and releasing oxygen as an incidental by-product.

Though no one seriously thinks that our supply of oxygen is in imminent danger of collapse, it is important to realize just where the daily replenishment of this most precious resource comes from. Most of the world's fresh supplies of oxygen are produced by single-celled, microscopic plantlike organisms floating near the surface of the oceans, supplemented, of course, by the photosynthesizing activities of terrestrial plants. The mighty oceans are the last great frontier of relatively un-despoiled natural habitat, but land-based human activity is beginning to sicken even them. Pollutants reaching the sea through streams and via the atmosphere (as gases are dissolved in water droplets), direct oceanic dumping, and the degradation of natural marine ecosystems through overfishing and mining operations are beginning to have their cumulative effects.

Consider what else green plants do for us. I was struck by a recent report detailing the salutary effects of a single, mature shade tree alongside a house in Chicago. Shade in the summer, insulation in the winter, and, amazingly, measurable purification of the air immediately surrounding the house. Once, while visiting the botanical gardens in Naples, Italy, a botanist told me that the air where we were standing was some six times purer than the air on the traffic-congested street only some 200 meters away from us! Green plants have the happy facility of filtering out noxious gases, utilizing carbon dioxide in the very act of photosynthesis, but also absorbing other noxious effluents and even particulate matter from dirty air. They give us life-sustaining oxygen and also act as filters—quite a dramatic bargain.

Plants do even more than enrich and cleanse the atmosphere. The Amazonian rain forest controls the water cycle in that region, as trees transpire a tremendous amount of water every day. In addition, tree roots hold soil in place, so that cutting forests always leads to massive amounts of erosion. Indeed, Earth is losing 25 billion tons of topsoil through erosion each year—a direct reflection of our conversion of natural vegetation for agricultural use, and an ominous portent of the difficulties that lie ahead for our continued reliance on agriculture.

Consider, for the moment, the net effect of increased erosion from denuded lands on the entire Biospheric system. Coral reefs which typically fringe the shorelines in tropical oceans are beginning to show worldwide distress. One factor in their decline is increased erosional runoff of silts from the cutting of tropical rain forests. For all their massive structure, corals are actually delicate colonial animals that are extremely sensitive to silt content in water. They quickly die when clear tropical waters become clouded with particles of clay and quartz, as has happened recently in Belize.

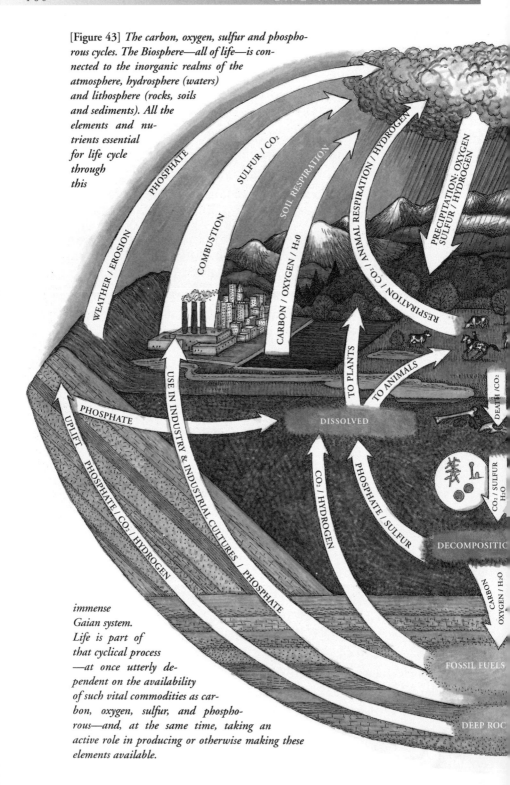

[Figure 43] *The carbon, oxygen, sulfur and phosphorous cycles. The Biosphere—all of life—is connected to the inorganic realms of the atmosphere, hydrosphere (waters) and lithosphere (rocks, soils and sediments). All the elements and nutrients essential for life cycle through this*

PHOSPHATE

SULFUR / CO_2

SOIL RESPIRATION

ANIMAL RESPIRATION / HYDROGEN

PRECIPITATION: OXYGEN SULFUR / HYDROGEN

WEATHER / EROSION

COMBUSTION

CARBON / OXYGEN / H_2O

RESPIRATION / CO_2

TO PLANTS

TO ANIMALS

DEATH / CO_2

UPLIFT

PHOSPHATE

USE IN INDUSTRY & INDUSTRIAL CULTURES / PHOSPHATE

PHOSPHATE / CO_2 / HYDROGEN

DISSOLVED

CO_2 / HYDROGEN

PHOSPHATE / SULFUR

CO_2 / SULFUR H_2O

DECOMPOSITIC

CARBON OXYGEN / H_2O

immense Gaian system. Life is part of that cyclical process —at once utterly dependent on the availability of such vital commodities as carbon, oxygen, sulfur, and phosphorous—and, at the same time, taking an active role in producing or otherwise making these elements available.

FOSSIL FUELS

DEEP ROC

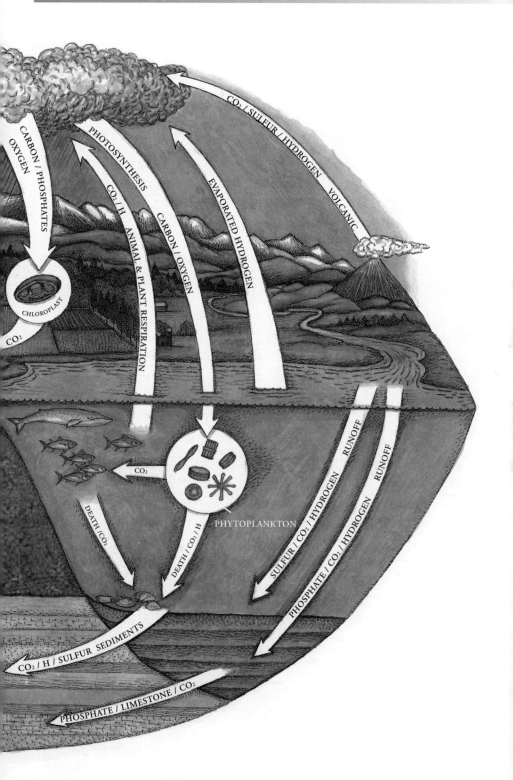

CARBON / OXYGEN

OXYGEN / PHOSPHATES

PHOTOSYNTHESIS

CO₂ / H ANIMAL & PLANT RESPIRATION

CARBON / OXYGEN

EVAPORATED HYDROGEN

CO₂ / SULFUR / HYDROGEN   VOLCANIC

CHLOROPLAST

CO₂

CO₂

PHYTOPLANKTON

DEATH / CO₂

DEATH / CO₂ / H

SULFUR / CO₂ / HYDROGEN   RUNOFF

PHOSPHATE / CO₂ / HYDROGEN   RUNOFF

CO₂ / H / SULFUR SEDIMENTS

PHOSPHATE / LIMESTONE / CO₂

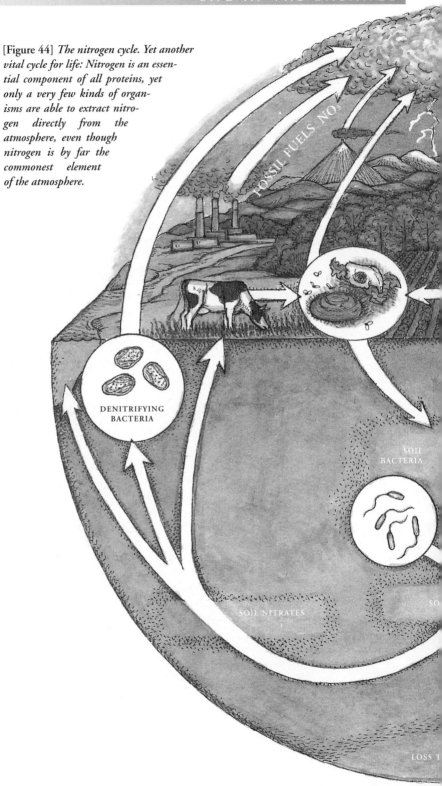

[Figure 44] *The nitrogen cycle. Yet another vital cycle for life: Nitrogen is an essential component of all proteins, yet only a very few kinds of organisms are able to extract nitrogen directly from the atmosphere, even though nitrogen is by far the commonest element of the atmosphere.*

FOSSIL FUELS NO₃

DENITRIFYING
BACTERIA

SOIL
BACTERIA

SOIL NITRATES

SO

LOSS T

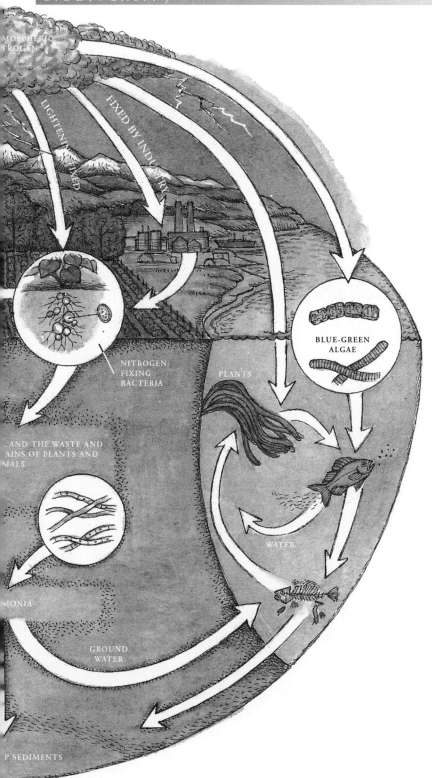

MOSPHERIC
TROGEN

LIGHTENING FIXED

FIXED BY INDUSTRY

NITROGEN-
FIXING
BACTERIA

PLANTS

BLUE-GREEN
ALGAE

AND THE WASTE AND
AINS OF PLANTS AND
MALS

WATER

MONIA

GROUND
WATER

P SEDIMENTS

Coral reefs themselves provide a protective barrier to other delicate ecosystems, such as the mangroves fringing the southern tip of Florida. As we have already seen, coastal wetlands are the breeding grounds of countless species of fish and shellfish. Without viable, functional coastal wetlands, our fisheries—already on the verge of collapse in many places, overtaxed as they are by incessant and often ruinous fishing practices—would soon be in even worse shape.

Everything is linked. We are accustomed to hearing that a complex web of energy flow—who eats whom—links all creatures in a local ecosystem. This is equally true of the Biosphere itself: What happens in the Amazonian rain forest ultimately affects not only the conspicuous mammals and birds of that forest, but also its fishes; runoff from the river affects the oceans, and all of its life. Decline of the fishing productivity on the Georges Banks off the Newfoundland coast can be traced in part to the cutting of tropical rain forests, as breeding of migratory fishes is disrupted far from the point where they are eventually hauled in by fishermen. The world truly is a complex system, and we are a part of it, still dependent on its renewable productivity, which we ourselves are beginning to stifle.

We have by no means exhausted the list of ecosystem services rendered by plants. Several absolutely essential elements—nitrogen and phosphorous to name but two—are derived from the plant world. Even though nitrogen is an essential element of all proteins, only a few forms of bacteria (aptly called nitrogen-fixing bacteria) can extract nitrogen from the atmosphere and incorporate it directly into their bodies. Some plants such as the legumes (peas and their relatives) harbor nitrogen-fixing bacteria amid their roots—a form of farming, in a sense. All the nitrogen in our bodies comes from eating plants or other animals that have eaten plants. The cycle is complete when decay organisms decompose dead plants and animals, producing ammonia, which is then converted by other bacteria to nitrates that plants can pick back up directly from the soil and to free atmospheric nitrogen. [Figure 44]

There can be no doubt that the Biospheric system—in particular, the vast range of organisms, from microbes to plants, fungi, and animals—plays a far greater role in our everyday lives than we think. We take them for granted—as we do our agriculturally produced foods, our cars, and our TV sets. That's fine—so long as we don't tip the applecart, by destroying so many of the world's local ecosystems that we compromise the Biosphere's cycles and our very existence.

A MORAL AND ESTHETIC IMPERATIVE Although we need the Biosphere's species for our own uses, and we rely on those species for the basic supplies of food, water, and chemical compounds on which all life depends, there is still a third category of concerns for the natural world—a third set of reasons to cherish the natural world and to resist its wanton destruction. Harder to define with precision, the esthetic appeal and moral challenge posed by Nature are to some the most compelling reasons to ward off the impending Sixth Extinction.

Not all of the world's major religions adopt the basic premise of the Judeo-Christian tradition, that the world and its living creatures were placed there by the Creator for our own human use. *Genesis* specifically exhorts humans to seek dominion over the beasts of the field. Recently, however, some theologians of the Judeo-Christian tradition have come to see the biblical exhortation as a call to stewardship; that is, our role ought to be as caretakers rather than as masters, to safe-guard the richness of the natural world, rather than to plunder it. Just as the *Genesis* role strikes me as an accurate assessment of the newly established order after the advent of the Agricultural Revolution (and *Genesis* was written not long thereafter), this newer theological interpretation, in my view, dovetails very nicely with the present condition of humanity vis-à-vis the natural world.

Other religious traditions, of course, espouse radically different views of nature and the place that humans take in the natural world. The oneness of humans with the rest of animate nature is perhaps especially apparent in the Hindu tradition, where the doctrine of reincarnation sees a continuity between humans and other species only matched in the Western world by the intellectual concept of organic evolution. Other traditions of the Far East—Buddhism and Shintoism, for example—also, in their different ways, locate humans as part of the natural firmament, and not especially exalted above the beasts of the field. These religions, it has been suggested, may make it easier for those raised in their tradition to see the urgent necessity of halting the blind, errant destruction of the natural world surrounding us all—easier, that is, than it is proving to be for those raised in Euro-centered, Western-world traditions.

Religion, of course, is not the sole source of moral suasion, and ethical concerns of what is right, and what we ought and ought not to do, have been arising on their own with increasing frequency as the early stages of the Sixth Extinction intensify. Cries for adopting a conservation ethic—one thinks here as much of Theodore Roosevelt and other far-sighted people of his generation as of the more recent (yet before their time) Aldo Leopold and Rachel Carson—are often tied to a sense of belonging to the natural world. A feeling, not just an intellectual grasp, of somehow still belonging to the natural world pervades the words of these early prophets—as when Rachel Carson, in the very title of her most famous book, asked us to consider what spring would be like without the songs of birds and the hum of insects.

These are essentially esthetic feelings—the notion that human beings just cannot expect to live completely successfully and happily strictly in the steel, concrete, and plastic world that increasingly appears to lie in the future. Famed Harvard biologist and biodiversity spokesman E. O. Wilson speaks of biophilia, his term for what he considers to be an innate sense of belonging to the natural world that, though subdued, is still present in all of humanity. Though we have proven to be mighty adaptable, I have a feeling he is basically right. The old saying "you can take the boy out of the country, but you can't take the country out of the boy" has a powerful analogue that encompasses us all. You can take people out of nature (local ecosystems, at least), but you cannot take nature out of people. That may well be the best reason for us to confront the Sixth Extinction.

SOME GLOBAL ISSUES "Study Finds Sperm Counts Are Declining." So reads a recent headline, not in the *Inquirer,* but in the *New York Times.* There is evidence—convincing to some, although not all, medical researchers—that the amount of sperm produced by human males in both Europe and North America has, on average, declined in the years since World War II. The story is dramatic, but the idea that someone could suggest such a far-flung effect on a biological function so fundamental to human existence is indicative of a much more general set of signals that the global system is hurting.

Scientists, thankfully, are a conservative lot, and all recent claims of global phenomena—including purely physical changes such as global warming, holes in the ozone layer, and increase in frequency and intensity of the El Niño climatic effect—have nearly as many doubters as proponents. It is always difficult to establish with absolute certainty that a recent trend—say, increase in global temperatures—represents the predicted effect of increase in carbon dioxide and other gases in the atmosphere that have been accumulating since the advent of the Industrial Revolution. After all, we have direct recordings of global temperature for at best only the last century. More importantly, we also know that Earth has been sporadically, but on the whole steadily, warming up *naturally* for the past 12,000 years, when the last ice age ended and the huge sheets of continental glacial ice began to retreat toward the North Pole.

It could be that the warming that has occurred during the twentieth century—a rise in global average temperature of 0.5°C causing a 2-centimeter rise in sealevel along the Eastern seaboard of the United States—is merely the continuation of a 12,000-year-old natural trend. On the other hand, carbon dioxide helps trap solar energy, keeping reflected light from bouncing entirely back into the voids of space. The equations are there: The more carbon dioxide in the atmosphere, the more solar energy will be trapped, and the higher the world's average temperatures will become. The recent dramatic rise in global temperature may well be, at least in part, a side effect of human activity: The progressive rise in burning of fossil fuels over the past several centuries and the more recent, but no less devastating, burning of forests, primarily in the world's tropical belts.

We may be uncertain and even skeptical, but we cannot afford to ignore the signs. The data are scant, so far, on the possible decline in human sperm counts, but there are several other apparent cases of negative global effects on life that seem almost certain to be caused by changes in the atmosphere—changes wrought as a purely unintended result of human activity.

One now well-studied example is the worldwide, and often rather precipitous, decline in frog and salamander populations over the last decade. I had noticed a sharp reduction in the half-dozen species of frogs and toads in and around my favorite little pond in the Adirondack Mountains of New York State. In the early 1970s, when I first started looking, there were green and bull frogs galore down in the pond, in populations so dense that the bulls were keeping us awake at night with their calls, and restaurateurs were shooting them for their bills of fare. American toads and wood and pickerel frogs were leaping all over the place along the pond's edge. Since the 1980s, it has been hard to see any of these species. But that is just one little spot, and for all I knew, I was just witnessing a natural fluctuation of population numbers—though it did strike me as odd that *all* the frogs and salamanders in my one little spot seemed to be declining at the same time.

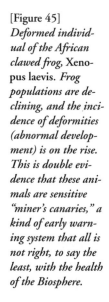

My Adirondack amphibian experiences are decidedly anecdotal—just casual impressions. Imagine my surprise when I learned that professional herpetologists had begun to notice an apparent worldwide decline in frog and salamander populations. More recently, frog populations all over the United States are producing many deformed individuals, a phenomenon first noted by Midwestern schoolchildren. [Figure 45] Just as in the case of global warming, experts on frogs and salamanders are divided on this apparent pattern. Is the decline real? If so, is it somehow just normal fluctuation, or is the decline at least in part caused by human-engendered environmental change? If human environmental disruption is the culprit, is the reduction of frog and salamander populations more a matter of alteration of local habitats, or is there some truly *global* factor at work? All herpetologists seem to agree that there is an urgent need for serious, long-term studies to get a firm handle on what precisely is happening to the world's frogs and salamanders.

[Figure 45]
Deformed individual of the African clawed frog, Xenopus laevis. *Frog populations are declining, and the incidence of deformities (abnormal development) is on the rise. This is double evidence that these animals are sensitive "miner's canaries," a kind of early warning system that all is not right, to say the least, with the health of the Biosphere.*

There is already enough evidence to convince some serious biologists that the amphibian decline is indeed real, and is affecting the great majority of the more than 5,000 frog and salamander species known to exist in the modern world. Severe alteration of local habitats lies at the very heart of extinction, and many frog populations seem to have been reduced through human conversion of their habitats for our own purposes. More subtle, but still in keeping with the theme of human interference on the local level, is the use of pesticides and the introduction of other toxicants. Back to the anecdotal, I do note that my observation of the frog and salamander decline in the Adirondacks in the 1980s coincided with a determined spraying program aimed at reducing the numbers of mosquitoes and black flies that attack humans and depress the flow of summer tourist dollars. Pesticides—or the human-induced drop in edible insects—may well underlie the reduction of amphibian numbers in many places.

Yet the tantalizing possibility remains that the worldwide decline may actually represent a truly *global* cause. In other words, the global pattern is not just the summation of isolated local effects but is actually caused by some factor that itself acts on a global basis. Frog and salamanders must return to the water to reproduce, and the skins of many species are delicate and more permeable to various substances than is the typical case for reptiles, birds, and mammals. Many species of frogs and salamanders are known to be sensitive environmental indicators; frogs, for example, are often good indicators of rising acid content in their native waters, as their numbers typically decline as levels of acidity rise.

What global factors could conceivably underlie the amphibian decline? Oregon State University herpetologist Andrew Blaustein—a cautious student of the problem—has suggested at least two possibilities. For one, there is a fungus known to attack, and to reduce the viability, of frog eggs. The fungus is found around the world, and perhaps a recent expansion of its range, or its potency, has contributed to the amphibian decline. More suggestive is the more recent hypothesis that an increase in the level of intensity of ultraviolet (UV) radiation is causing the decline. Blaustein and his colleagues have found that the developing eggs of frog species known to be declining are more sensitive to—more damaged by—exposure to a given level of UV radiation than are the eggs of frogs whose population numbers have remained relatively stable in recent years.

Ultraviolet radiation. That rings a bell—as anyone who has read of the recent alarming rise in incidence of human skin cancer well knows. The medical literature has established a firm link between exposure of human skin to the sun and the occurrence of skin cancer—a correlation that, like that between lung cancer and smoking, implies actual cause and effect to most medical professionals. Dermatologists are in virtually unanimous agreement that exposure to UV radiation is one of the causes of skin cancer.

Careful measurements by atmospheric physicists indicate that uv radiation has definitely been on the rise. The increase is especially apparent in the higher latitudes, closer to the North and South Poles than to the equator. Much of the UV component of solar radiation is absorbed by ozone, a molecule consisting of three atoms of oxygen. Each winter, the natural layer of ozone in the atmosphere thins, at places disappearing entirely. In recent decades, satellite imagery and balloon probes alike have revealed increasingly large and persistent rents—holes—in the ozone layer. Damage to Earth's protective ozone layer was actually predicted: Chlorofluorocarbons—organic compounds widely and routinely used in aerosol cans after World War II—were known to interact with ozone, destroying large quantities of it in a one-way chemical reaction. That, it is widely agreed, is precisely what has happened. Spray cans, in this roundabout yet deadly way, are responsible for the epidemic increase in human skin cancers and perhaps the decline of many amphibian populations as well.

There are several critical lessons here. First, we see that humans are indeed capable of altering the global system—in this instance, the atmosphere. We also see that such changes can have profound effects on living things—in this particular example, most clearly and convincingly documented in the rise of skin cancer in humans. We also see that negative effects—amphibian population declines—occurring globally may result from a single global cause, or may be the simple, yet nonetheless devastating cumulative effect of local habitat disruption and destruction.

There are other lessons as well. Though some biologists point to the economic importance to humans represented by frogs (by keeping insect pest populations at bay, for example), and though many of us feel the noisy booming of bull frogs is a welcome nighttime experience, many citizens of the modern world more than likely would maintain that the world (meaning, of course, the *human* world) could get along just fine without those 5,000-odd species of frogs and salamanders. They would be missing the main lesson here: The real significance of the global decline of frogs and salamanders is that the amphibians are telling us something about the state of the atmosphere and, thus, of the global system as a whole. Human skin and frogs eggs just happen to be at the higher end of sensitivity to UV radiation. Amphibians—and human skin for that matter—are global analogues of miner's canaries, early warning systems that all is not right with the global system.

The emerging story of the increase of atmospheric UV radiation through human agency—and its boomerang effects on us and quite likely all other living systems—is stark, dramatic, and fairly easy to comprehend. More subtle are the negative effects on human life of human-engendered destruction of natural ecosystems and the consequent loss of species. It has proven difficult to explain what the loss of the northern spotted owl of the old-growth forests of the Pacific Northwest really means to local people, especially those engaged in the lumber industry in Oregon and Washington, to those of us living elsewhere in the United States, and ultimately to the people of the world. It simply doesn't make much intuitive sense to claim that the loss of a few hundred pairs of owls—which hardly anyone actually ever sees—will have a profound negative effect on human existence. Yet it does matter. Those owls are a litmus test of the health of the ecosystem in which they live, and we are now beginning to understand that we depend on those systems far more than we imagined since we invented agriculture and stepped away from the local ecosystem.

THE SIXTH EXTINCTION Biologists do not know
exactly how many species are currently on the planet. Science has recorded and named some 1.6 million species, but we know this can only be some small fraction—no more than 10% to 15%—of the true number. Some biologists believe we have identified only 1% to 3%, and that there may be as many as 100 million species on the planet. Because concern over the accelerating loss of species has been mounting, biologists have turned in earnest to the key question, How many species are on Earth? They have begun to converge on an estimate of some 14 million species, but opinions still sharply differ on this vital issue.

Why are we so ignorant of the biotic riches of Earth? Scientific survey of the world's species began in the seventeenth century, but did not switch into high gear until the mid-nineteenth century, when the heyday of European colonialism mixed with the Industrial Revolution, producing a blossoming of exploration and scientific inquiry. Naturalists like Alfred Russell Wallace and Henry Walter Bates traveled to then-exotic destinations such as the Amazon Basin and the Spice Islands (part of present-day Indonesia) to amass vast collections of plants, insects, spiders, aquatic invertebrates, fish, amphibians, reptiles, birds, and mammals. The collections of such as intrepid explorers found their way at first into privately held "cabinets" of natural history and increasingly into the large natural history museums that were founded in the mid-nineteenth century—museums such as the British Museum of Natural History in London, the Natural History Museum of the Smithsonian Institution in Washington, D.C., and my own favorite treasure trove, the American Museum of Natural History in New York. Natural history museums are libraries of biodiversity, storehouses of the world's biological riches, where scientists can compare specimens and assess the identities and evolutionary relationships of the world's species.

Though research biologists at major universities have historically contributed to the effort of finding, describing, and naming the world's species, increasingly this role has concentrated in the hands of the scientific staff of major natural history museums. *Systematics* is the branch of biology devoted to describing Earth's species, analyzing their evolutionary relationships, and producing biological classifications. Because biology keeps expanding (most recently into the realm of biomolecules), and because vast collections of specimens are needed as part of the routine work of systematists, museums have become *the* focal point for systematics research.

Needless to say, there are far more species than experts to identify them. For some groups, there are few (sometimes no) experts actively working to inventory the world's stock. One way biologists frame accurate guesstimates of how many species probably exist in the world is to assess what we know we have found already, observe the rate that new species are turning up, evaluate how concerted the effort is to find new species for a given group, and derive some sense of what might still be out there, as yet unidentified.

Ornithologists think that they have found most of the world's bird species, as the number has begun to level off at around 9,000, and mammalogists also think they have described and named well over 90% of the world's mammal species. As we have seen, even large species of mammals still turn up on a regular basis, such as the large antelope and deer recently discovered in recently war-ravaged Vietnam and the several species of lemur discovered over the past decade on Madagascar. New bird species also turn up regularly. Because so many systematists have focused on birds and mammals—big and obvious, the charismatic megafauna—and because much more numerous groups, such as insects, have received relatively much less attention, the ratio of named to as-yet-undiscovered species varies widely from group to group.

There is yet another major source of inference for assessing the actual number of species on Earth, one that is tied into the very critical question, How do we know that we are in the midst of a sixth, major global mass extinction? The connecting link is *habitat*, by now familiar as the essential ingredient in species loss, but one which, quite obviously, underlies the sheer existence of species.

In an elegant series of studies, Smithsonian coelopterist Terry Erwin came to his by-now famous—and still controversial—estimate that there are some 30 million species of insects in the Tropics alone. Erwin carefully sampled the insect (especially beetle) faunas of various forest canopies in Panama, Brazil, and Amazonian Peru. Erwin's goal was to determine how limited beetle species were, on average, to particular types of trees. Then, taking into account the total aerial extent and canopy diversity of the tropical rain forest, he was able to derive an estimate of the total number of beetle species currently in existence, and from that estimate extrapolated an estimate for the total number of insects.

Coming as it did when most of the world's biologists were still thinking that there are at most only a few million species on Earth, Erwin's analysis shocked a lot of biologists into attention. More recent work, including similar in-depth, total assaying of a region's biotic riches, have tempered his estimates somewhat, but we now are accustomed to thinking that there are at least 10 million species of insects, rather than 1 to 2 million species, a 10-fold increase in our estimates of Earth's living species diversity. If so rich a percentage of the world's species is yet undiscovered, then their loss, their undocumented extinction, becomes more critical. Loss of unstudied genetic diversity means never to have that knowledge and never to be able to utilize whatever potential those species might have had, one of the major reasons humanity has become alarmed at the growing rate of destruction of the world's species.

How do we measure that rate of disappearance? One simple and obvious way is to record the number of species—such as the passenger pigeon, the dodo, and the great auk—that are known to have become extinct in historic times. A related approach is to evaluate how well species that have been placed on threatened and endangered lists have fared. Appendix I shows the documented species extinctions since 1600. Although they are considerable, they seem hardly the stuff of catastrophic mass extinction. Is the current loss of biodiversity overestimated? Are we really in the throes of a major mass extinction, the Sixth Extinction?

Other approaches to measuring extinction rates yield more alarming results. Many species in museum collections simply can no longer be found in the wild. Many of my colleagues at the American Museum of Natural History have told me of returning to locales where, a few years earlier, they had found new species—in one case, a new species of spider in Chile—only to discover that the species' habitat had disappeared.

Thus, we can use the very same sort of reasoning that gives us one way of estimating the number of species that exist (complete sampling of species richness in finite areas) to yield an estimate of the rate of species loss. We can measure the rate of habitat destruction. Aerial and satellite photography have revealed that tropical rain forests are being destroyed at a rate of some 12 million hectares a year.

Combining known rates of habitat destruction with accurate assessment of numbers of species occurring per hectare in tropical rain forests, and the average size of areas they occupy, yields rough but credible estimates of current rates of species loss. We have already encountered the most famous of these: E. O. Wilson's figure of 27,000 species a year, which boils down to 3 species an hour lost forever. Some biologists think Wilson's figure is too high, but plenty more think he has underestimated the situation.

How can we reconcile the rather low rates of extinction revealed from actual documented examples with the vastly more impressive—and troubling—estimates based on loss of habitat acreage? Most of the documented examples of extinction in the past few hundred years involve large, easily observed species (such as the quagga, a zebralike horse species); species from islands, where diversity is most easily monitored (the dodo was a giant, flightless pigeon living on the island of Mauritius in the Indian Ocean); and the well-studied north temperate latitudes, where the vast majority of the world's biologists have lived and worked (the passenger pigeon was a North American species). We know about these species simply because we have had the opportunity to study them and observe their loss.

There is more to the story than simple convenience of study. The Tropics harbor vastly more species per acre than the higher latitudes, where species are physiologically broadly adapted, and consequently distributed over far greater ranges. Thus, a species with huge populations spread out over a large area can suffer huge reductions in numbers, yet still be nowhere near total annihilation. Tropical species are much more narrowly adapted and narrowly restricted in geographic range so that the same amount of habitat destruction in the Tropics accounts for many times more species extinctions than would occur in the higher latitudes.

The conclusion seems clear: Ongoing, unrelenting habitat destruction is driving thousands of species to extinction each year—species that for the most part we haven't even come to know as yet. Direct destruction of ecosystems is having a mounting effect on ecosystem services, which are vital to our quality of life and our continued ability to elude extinction.

Two additional aspects of human disturbance have led directly to species extinctions and current endangerment. One is simple overexploitation—overhunting and overfishing. The great auk, the passenger pigeon, and many other species were simply hunted to death, and modern technology has made marine fishing wasteful and deadly.

The other aspect of human interference with the global ecosystem is, for the most part, nonintentional. It is, as well, something of a reversal of the generalization that extinction always *follows* ecosystem disruption and degradation; in this instance, the extinction of species *causes* the disruption—or so it seems when species (domesticated and wild, plant and animal) follow humans around the planet. Many species have been deliberately introduced, as when European colonialists brought samples of domesticated and wild species with them for utilitarian and, presumably, nostalgic and esthetic reasons. Introduced species can and often do drive individual native species directly to extinction, and their loss sends shock waves through the ecosystem. So common are European birds that it is difficult to observe native birds in New Zealand except in the remoter regions. New York City's Central Park, so justly famed for being home to well over 200 migrant and permanent resident species, is nonetheless dominated by starlings, house sparrows, and pigeons—all three European imports. House finches—recent immigrants from the West Coast—are also on the rise.

Thus, there is a definite symmetry, a balanced give-and-take between the players in the game of life and the ecosystems in which that game is played. If ecosystem degradation is the basic reason that species are lost to extinction, sometimes the players—by driving others extinct—are the ones disrupting the ecosystem. Though it seems likely that such events have happened in the deep geologic past—as when North and South America became joined, resulting in migrations and the loss of many mammalian species—we must bear in mind that the present-day, almost chaotic dispersal of alien species around the globe is yet another manifestation of the disrupting hand of humanity and the destructive effects we have wrought on local ecosystems the world over.

We began our course of negative impact when our species *Homo sapiens* began spreading around the world in the last 100,000 years. We intensified that impact with the invention of agriculture and our concomitant declaration of virtual war on local ecosystems. Our impact as been accelerating, exacerbated by the Industrial Revolution and, above all, by the twin phenomena of runaway population growth and the unequal distribution of wealth and consumption patterns.

Chapter 6 STRIKING A BALANCE   The Panama Canal is an

engineering marvel. Opened for traffic in August of 1914, at

the cost of tens of thousands of lives and hundreds of millions

of dollars, the Canal has earned back roughly two-thirds of

the $3 billion dollars invested in it since 1903. It costs larger

ships up to $200,000 to make the 80-kilometer voyage be-

tween the Atlantic and Pacific Oceans, a trip that takes about

9 hours to complete. The real point of the Panama Canal, of

course, is the tremendous savings in time, fuel, and money this

ditch across the narrowest sector of the Isthmus of Panama of-

fers human commerce. To round Cape Horn at the very bot-

tom of South America adds 3 weeks to a voyage. ❧ What one

makes of the Panama Canal depends very much on one's point

of view. On the one hand, as a consumer of goods, the average American profits enormously from the Canal's existence: Prices, bad as they may be, are simply not as high as they otherwise would be for a New Yorker buying a Japanese car, or a Californian drinking Brazilian coffee. On the other hand, the Canal was ripped right across a stretch of lush tropical rain forest; huge landslides further destroyed the terrain as sections of hillside along the continental divide collapsed into the ditch while it was being dug, and even after the canal had opened to shipping. The huge artificial lakes that are so crucial to running the Canal claimed vast tracts of rain forest. It is a distinct trade-off, perhaps even a mixed blessing this Panama Canal: In saving time and money, it accelerates the despoliation of the world's ecosystems (as when, for example, exotic woods are shipped through the Canal)—in addition to the initial (and ongoing) destruction of Panamanian rain forest posed by the very existence of the Canal.

Yet, on the other hand, saving fuel is an environmental plus. Even that is not all, for the very operation of the Panama Canal is an object lesson in tradeoffs. And the most fundamental lesson of all: Somehow humanity must once again learn to live in synch with the environment, in partnership and stewardship rather than exploitatively with the natural world. It turns out the Panama Canal, initial destroyer of rain forest, absolutely depends on the health and welfare of the remaining rain forest still surrounding it simply for it to continue to function.

Early plans for the Panama Canal envisioned a sealevel pathway, without the necessity of a complex system of locks to raise and lower the ships as they crossed the continental divide. A sealevel canal would have brought far more environmental problems than posed by the actual canal, in which ships are raised and lowered a total of 26 meters through a system of 5 locks. The Isthmus of Panama was only completely raised above sealevel some 3 million years ago. Simultaneously, North and South America were connected by a complete land bridge, allowing interchange between northern and southern species, and disconnecting the waters of the Atlantic and Pacific. The result was extinction of many land species as invaders from both south and north expanded their terrestrial ranges, and, at the same time, speciation—true evolution—of marine creatures, as Atlantic and Pacific species, now isolated, quickly went their own evolutionary way. A sealevel canal would have reversed the situation—at least for the marine species—and much extinction would surely have occurred on both sides of the Isthmus. As it happens, the Panama Canal is largely freshwater, and the barrier between Pacific and Atlantic is maintained. Why would human-caused extinctions pose greater problems than the extinctions caused through geologic events, such as raising the Isthmus of Panama? Human disruption of ecosystems tends to occur very rapidly, and the net effects of destabilizing many ecosystems around the world—and the extinctions of species that inevitably ensue—are cumulative, and are beginning to threaten the quality of life for humans as well.

The Panama Canal is not a sealevel ditch at all; rather, it is an ingenious system of watery conveyance that seems for all the world like the fabled perpetual motion machines of old. The Canal, in effect, runs off the very rain forest it cuts through. It takes some 220 million liters of water to flush a ship from one ocean to the other, to raise and lower each ship 26 meters through the five-lock system. The critical step was damming the Chagres River, which originally ran from the central highlands of the Isthmus to the Atlantic Ocean. The upper dam created Madden Lake, while the lower dam, located not far from the Gatun Locks (a double system of locks on the Atlantic side) created what was, when first built, the largest man-made lake in the world: Gatun Lake. Water flows down from Gatun Lake through the Gatun Locks on the Atlantic side, and toward the Pacific, down through the single Pedro Miguel locks to Miraflores Lake, through the double Miraflores locks, and out into the Pacific Ocean.

In other words, the system runs on gravity. Even the electricity that powers the motors that open and shut the massive steel doors (the originals built in Pittsburgh are still in use!) is generated as hydroelectric power as water spills through the dams. (Testimony to the marvelous engineering effort of the Canal, these huge steel doors are open and shut by 40-horsepower electric motors!)

But one question remains, and the answer to it shows that the Panama Canal is in reality no perpetual motion machine: Where does the water come from? Two hundred and twenty million liters times the approximately three-quarter million ships that have passed through the locks is a lot of water. Only once in the long history of operations has the Canal not had enough water to stay in operation. Something keeps replenishing the supply—and that something is, of course, rain. Lots of rain. But why does it rain so much in Panama? Because Panama *is* a rain forest—at least where the forest has not been destroyed for cities, agriculture, and, of course, the Canal itself.

Plants—especially the trees and other green plants of the tropical rain forest—regulate atmospheric gases like oxygen, carbon dioxide, and water vapor. During the rainy season, everyone who has visited the Tropics knows that dense cumulus clouds begin to coalesce, darken, and lower until an intense cloudburst saturates the region. The clouds begin to dissipate, and all is clear at nightfall. Day after day, the afternoon rains come, and the water they bring is largely supplied from the leaves of the trees of the forest, trees whose roots eagerly suck up the rain as it falls.

In short, no forest, no reliable rainfall, and, in Panama, not enough reliable rainfall to run the mammoth Panama Canal operation. Panamanians, who are scheduled to take control of the Canal completely in the year 2000, have recently become aware of just how fragile the system is, just how dependent on rainfall the Canal is, and how dependent the rainfall is on the rain forest. Fortunately for the operations of the Canal, much of the original rain forest on the edges of the Canal remains pretty much intact, and the Panamanian government is actively seeking to create a system of reserves and refuges as unbroken as possible along the Canal's periphery.

The moral of the story: The Canal is a reality, a vital component of the daily $1 trillion global economic exchange among humans. Few expect or desire that international trade cease or even diminish appreciably: The die is cast, and humanity can neither afford, nor is able, to retreat to some more insular, pristine state. At the same time, humans must stop taking for granted the natural resources—Earth's land, mineral and petroleum wealth, freshwaters, and animal, plant, fungal and microbial species. The free market (and the economists who study them) must factor in the cost of depletion of commodities so vital to the ongoing quality of human life. The whole world is full of examples, if not so vivid, then at least as starkly true, as that of the Panama Canal. To continue working, we must see the partnership that we have (but have so badly abused) with the natural world. We must work with the natural world—as the Canal engineers work with the rain forest—to keep the system going. Above all else, we must develop a keen sense and practice of *sustainability*, where we use the resources of the natural world in a manner that allows recovery (if such is possible: fisheries will rejuvenate if the pressure of overfishing is removed, but oil and metallic ore deposits are finite and non-renewable). We must strike a balance, and learn to live in equilibrium with the natural world around us.

WHAT CAN WE DO? We have seen that the world is in the midst of a new surge of extinction—one that is taking some 27,000 species a year from the planet, many of which are doomed before anyone has the opportunity to find them, to study them, to name them, and perhaps to realize their worth. The living world is valuable to us in many different ways. For our foods and medicines, in particular, we continue to rely on the wild. Even our domesticated animal and plant species—whose genetic stocks are rapidly being homogenized and thus depleted—need wild ancestors and collateral kin to provide added genetic variation, to improve or simply to save them.

We know that, by ceasing to live within local ecosystems, we have become estranged from the natural world. We have largely lost sight of the fact that human life still depends on abundant supplies of freshwater, nutrient-rich soils, and clean air with the right balance of oxygen and other gases. Here is the message of the Panama Canal: Ecosystem services are absolutely vital to human life, and though we tend to take them for granted, loss of topsoil and diminishing availability of sufficient supplies of clean water are like lit fuses racing down to the powder keg. Political problems, not only in Third World developing nations, but also in developed nations such as the United States of America, are bound to intensify as these problems are left unaddressed. They are doing so already. All of these issues are very much bound up with the survival of the living world in some form—if not "pristine"—at least in some recognizable facsimile of "natural."

There are also the harder-to-quantify reasons for stemming the tide of the Sixth Extinction: ethical and esthetic reasons. Moral reasons. To drive a species like the giant panda to extinction is, in my book at least, to commit a form of genocide. The habitat of the giant panda has shrunk to a tiny fraction of what it was at the turn of the last century. Most of their original habitat has already been converted to farmland to feed and fuel the burgeoning growth of the Chinese people. No one—least of all I—would argue that children must starve to save the last remnants of pandas in the wild. But the system of reserves set in place (barely just in time), with the help of such eminent scientists as George Schaller, to save the last patches of giant panda habitat, does imply that the local Chinese will have to make do with existing lands under cultivation—rather than convert what little remains of panda habitat to farmland. After all, if those patches were co-opted as farmland, it would soon seem still not to be enough—as the spiral of food production and population continues. As we shall see below, viewing pandas—icons as they are of endangered species—as economic assets, rather than as liabilities, is part of the solution.

We humans are, originally at least (the past 10,000 years of agriculturally based existence notwithstanding) of and from the natural world. However divorced from the wild we city dwellers might be, there is little doubt that nearly all of us experience some feeling of peace, joy, and downright comfortable familiarity when we are exposed to wilder places, even if it is just a park within the confines of a bustling city.

We have seen, too, that the Sixth Extinction is very real. We know it comes about through human intervention with the wild. Humans are still converting terrestrial ecosystems to farmland at an unprecedented clip (one estimate has a half hectare of tropical rain forest disappearing every second!), and, to a lesser extent, ecosystems are disappearing under suburban housing developments, shopping malls, highways, and eternally expanding cities. Add to that the devastation of overharvesting (trees and fisheries, in particular) and the damage caused as we introduce alien species around the world. This was staved off when the decision was made not to construct the Panama Canal at sealevel. As we have already seen, the devastation is a constant and pernicious side effect of ongoing human contact around the globe.

In other words, we have a problem, and we know what it is. We *understand* that we ourselves are causing the problem, so what are we going to do about it? Here is my agenda.

1. *We Must Acknowledge the Problem.* Because the vast majority of the world's nearly 6 billion people no longer live inside whatever remains of their local ecosystem, we must begin a systematic campaign to enlighten the world's population—rich and poor nations alike—to the

dilemma. We are faced with a massive public relations problem. This point re-emerges below, in conjunction with sustainable harvesting, ecotourism, and other issues in poor, developing nations. But the issues need hammering home just as much in developed nations as in the Third World. Most economists in the United States, for example, still fail to acknowledge the problem, with some (such as University of Maryland's Julian Simon) actually denying that species loss really matters to human life. Simon also believes that we can never have too many people living on Earth, which takes us to the next item in the agenda.

2. *We Must Stabilize Human Population.* Alteration and destruction of the world's ecosystems has accelerated in the past century and is a direct reflection of the virtually out-of-control growth in human population. The problem is only exacerbated by the unequal consumption of resources: As we have seen, Americans have a much greater impact on the planet, per capita, than do citizens of Third World nations, as if our population were many times greater than it actually is. Thus, the population problem pertains to developed nations every bit as much as it does to undeveloped countries where the absolute numbers of humans may be very much greater.

Yet, environmental woes and the Sixth Extinction are not just the effects of rich nations removing the expensive commodities from the poorer countries. Poverty itself leads to an ever-widening circle of environmental annihilation, as people desperate for enough firewood to cook their meager evening meal, and to shed a little light and warmth are just as responsible as foreign lumber companies for the destruction of the world's forests.

What, then, can be done to stabilize world population? Population biologist Joel Cohen, in his provocative analysis, entitled simply *How Many People Can the Earth Support?* (1995), agrees with pundits that human population will eventually stabilize of its own accord. Why? I believe it is because we humans truly have redefined our niche, becoming the first species to play a direct and concerted economic role within the global system. In addition, economic systems have carrying capacities. Just as only so many squirrels can live in New York's Central Park—limited as the food supply is—only so many people can inhabit Earth.

How many people is that? Joel Cohen says, wisely, that it depends at what level of support you are talking about. If everyone were to live at the level of comfort of the average U.S. citizen (begging the question of extremes of poverty and wealth *within* the United States—we are talking averages for the moment), we have undoubtedly passed the maximum number of people Earth can support. One or two billion, according to most estimates, are the most Earth can support in the fashion of a middle-class American. Accepting the disparity in wealth between nations, accepting, that is, that the vast majority of the world's population currently living in poverty will continue to do so, how many people can Earth support? Twelve billion? Fourteen billion? That's the ballpark range of most estimates of when the stabilizing system will automatically kick in. Long before those kinds of numbers are reached, regional and interregional conflicts, famine, and disease threaten to make recent events around the world pale by comparison.

We simply must stabilize human population growth before it is stabilized by forces out of our control. How do we do that? Economists and demographers, pointing to the typically slower rates of population growth in industrialized nations, have long championed the idea that economic development, through industrialization and other means, will work everywhere it is applied. Make people richer by improving their economies, and they will have far fewer babies. In a very real sense, this is probably quite true. Yet it should be obvious to anyone who will look that, as a cure-all, global economic development is sheer fantasy, a pipe dream. There is no way the standard of living of a place like Bangladesh will be raised to anything even remotely like current U.S. standards. Grim as it sounds, it is more likely that U.S. standards will (continue to) fall than Bangladeshi standards will significantly rise.

Are there no other ways to stem the still-rising tide of human population? Cohen lists several, none of them easy; he recommends trying as many as possible. Dissemination of birth-control information and paraphernalia is one obvious avenue and has had good results. Economic development, where it can be achieved, always seems to help. The ideas that have attracted the most attention at recent international Congresses on population and the status of women converge on the *education and economic empowerment of women in the context of their own local cultures* as perhaps the most promising approach of all. Studies and pilot projects in several developing nations have shown that the birthrate drops significantly in proportion to the number of women who become educated and who hold positions in the local economy. Once again, education is the key: learning that there is a problem, on the one hand; learning about birth control; but also learning how to take an active role in economic affairs outside the home, a path that seems automatically to dampen birthrates.

The latest figures that I have seen show the world's population growing a bit more slowly than had been projected—but we are still adding nearly a quarter of a million people *every day*. That is simply too many for our own human good and certainly too many if our planet, and our planet's ecosystems and species, are to survive in anything like recognizable form.

3. We Must Rewrite Economics Texts and Fine-Tune the Notion of Sustainability. As we have seen, local populations of absolutely all species other than our own live within, and are limited by, the productivity of their own local ecosystems. They are a functional, dynamic *part* of that local ecosystem. Organisms live healthily and well within those systems, except when the systems suddenly become less productive (through drought, for example, or other prolonged climatic change). Human beings appear already to have exceeded the carrying capacity of the global system—if we choose middle-class U.S. standards as living "healthily and well" (itself of course a debatable point).

In any case, living in a world where most of its human denizens are living in poverty (decidedly not healthily and well) has enabled us to exceed the lower estimate of how many people Earth can support. We have no sense of equilibrium, no sense of "enough"—whether it be consumer items (including food), or the number of people everyone can tolerate crowding the planet. To stabilize, we must break the cycle of endlessly expanding farmland. (There is, after all, a finite amount of land. If the Chinese farmers had been allowed to wrest the last remnant acreage of usable habitat from giant pandas, they would soon have found out that there was no more agriculturally usable land to take. The panda would have become extinct, and there would still be the dilemma of finding more land to raise crops.) We must break the cycle of "we have more mouths, we need more food; we have more food, but now we have more mouths so we need more food."

The goal must be to raise the quality of living of every one now living on Earth, not to produce more souls to share less and less. This is the essence of *sustainability*. We need to raise, to mine, to quarry, to make only what we need to sustain ourselves in our present numbers. We need to husband our nonrenewable resources. We need to recycle. We need industry and economists to stop treating natural resources as if they were God-given gifts intended only for our use. Economists must begin to reckon, not just with depletion, but with the impact, the *cost*, on the local and global systems of the removal of such resources. We need to utilize resources everywhere in a fashion that will sustain ourselves but replenish renewable resources, so that responsible harvesting will leave plenty for the next generation—which ideally should be no larger than the present generation.

How do we do this? How do we develop a sense of "enough"? See the previous two points—but see, also, the next agenda item. As always, we need to become aware of the problem. We need to educate, if there is to be any hope of self-reliance in facing our ecological future.

4. *We Must Utilize Our Existing Expertise in Conservation.* If it is true (and it most certainly is!) that we do not have an accurate grasp of how many species are currently living on planet Earth, we do know that Earth's ecosystems and species are disappearing at a horrifyingly rapid rate. What can we do to halt the flood of habitat disruption and extinction?

A profound change has taken over both the field of conservation biology and the practice of conservation in the field. Nearly everyone now agrees that habitat loss and ecosystem degradation are the main culprits of the Sixth Extinction and therefore that conservation efforts should be aimed first and foremost at *identifying and saving critical habitat areas.* This does not of course mean that we should never focus attention on individual species. After all, often it amounts to the same thing. The outcry to save the giant panda or the northern spotted owl *meant* saving as much as was left of its traditional habitat. Indeed, focusing on individual species helps conservation biologists identify which habitats and ecosystems are most urgently in need of protection.

Saving stretches of habitat, or significant portions of ecosystems, all boils down to one thing: conserving acreage in pristine, or as-near-as-possible pristine, conditions. How do we identify such areas? How do we determine how large, or what shape, such areas might take? Here, a focus on species actually can help. For example, some biologists in the United States have recently identified hot spots, concentrations of rare and endangered species, which they claim are sufficiently small in extent that they can be readily conserved. Other biologists focus on *areas of endemism*, pockets where a number of species unique to a region are concentrated, such as the island of Madagascar. A further example comes from my colleague at the American Museum of Natural History, Melanie Stiassny, who has pointed out that the two species of cichlid fishes in Madagascar are very primitive; the hundreds of cichlid species swarming in the lakes of the East African Rift Valley system are for the most part all recently evolved. Perhaps, Stiassny suggests, the two primitive species in Madagascar should take priority over the African species (threatened as they truly themselves are), as the Madagascar species uniquely preserve an aspect of the early evolutionary diversity of the group.

All such suggestions have merit (though I am concerned that the "hot spots" idea for the United States minimizes the sheer extent of habitat that we ought to be conserving). But comparing ecosystems and component species to decide what to save and what to let go is a truly tricky proposition. Is one ecosystem more valuable than another because it has more species in it? If so, then tropical systems are by definition more valuable than their higher-latitude counterparts. Are we to sacrifice the tundra because it has far fewer species than a tropical rain forest?

Clearly, making decisions on preservation must hinge on factors that transcend the presence of this or that particular species—or even the number of species at risk, or their uniqueness, or whatever other criterion we might adopt. If we take the longer view, if we realize that it is intact ecosystems themselves, and not just their component species, which have value—with ecosystems services and esthetics ranking the highest—then we shall realize that all ecosystems, all habitats, all natural places, have such value, and that as much as is practicable deserves to be, must be conserved. As much, that is, as can also admit to sustainable use and the economic well-being of local human inhabitants.

5. *We Must Strike a Balance between Human Economic Needs and the Continued Healthy Existence of Ecosystems and Species.* It is simply no good to tell the head of a starving family in Madagascar not to

cut down trees for fire wood, or not to burn the forest to clear land for more rice cultivation. Such postures are worthless, arrogant, simplistic, and doomed to failure. The more enlightened conservation activists have come to realize that the "armed camp" approach—where fences are erected around tracts of forest or savannas, and the locals told to keep out—simply won't work. *Sustainability* refers as much to human life as it does to the use of local environment. Conservation, in short, is doomed unless the economic interests and well-being of local peoples is taken into central consideration. This applies to the spotted owl/logging quagmire of the Pacific Northwest every bit as much as it applies to lemur reserves in the Madagascar rain forest.

If "Just say no" is not appropriate and, in any case, doesn't work, how then to incorporate the economic needs of local citizenry with the equally vital need of stopping the outright destruction of the ecosystems around them? If it is true that wholesale prohibition of logging would throw much of the Pacific Northwest into a potentially devastating economic tailspin, according to some estimates, it is also true that, left unchecked, all the trees will be cut—all the logs collected—within a decade anyway, merely forestalling the economic debacle for a short-sighted few years. The time, obviously, has come for diversification in that particular lumbering economy before an economic debacle comes from *either* complete prohibition or simply cutting down the last tree.

People may be divorced in a formal, ecological sense, from their local ecosystems, but, especially in developing lands, they still freely use the fruits of their local ecosystem. They take animals for food and pelts; they cut trees for building and firewood. They have a sense that it is their right to do so—and, of course, in so thinking they are perfectly correct. If that attitude underlies the ongoing destruction of the environment around them, it also, ironically enough, holds the key to persuading local people the world over to adopt more of a stewardship role, to strike a balance with their environment. They must be educated to see that the living world around them is more valuable to them *alive* than *dead*. That, for example, was precisely what American biologist George Schaller and his colleagues were able to do vis-à-vis the Chinese governmental attitude toward the giant panda.

How, then, to convince a young Malagasy man that Madagascar's lemurs must be saved, and that the only way to do so is *not* to cut down the woodlands right around his home? The answer is an influx of money from the outside. Not outright gifts or bribes. Rather, it is necessary to establish economic enterprises that make it possible, not just for isolated individuals, but for entire whole communities to see the worth, the value of the living world around them. Value not in a traditional sense—as firewood, say—but new forms of value.

Costa Rica is justly famous for its system of national parks established to preserve what was left of a rapidly diminishing rain forest (much of which was destroyed for the planting of coffee, which grows primarily between 1,160 and 1,600 meters above sea level). It is also renowned for its INBIO program, where the pharmaceutical firm Merck paid the Costa Rican government $1 billion for prospecting rights for new, potentially useful (and marketable!) drug compounds. Oddly enough (at least in my book), molecular biologists (or, more usually, their university or industrial employers) are allowed to *patent* the gene sequences that are the instructional codes for the production of such natural biological molecules. If one could argue reasonably that no such snippet of genetic code belongs to anyone in any case, if such information can be said to "belong" to anyone, it should not belong to the people who find it, but to the people on whose land the plants or animals with this useful information actually live. The INBIO program recognizes precisely this point and dramatically helps Costa Ricans understand that their rain forests have value beyond a simple source of wood, or, worse, cover that must be removed to plant still more coffee.

Costa Rica has led the way in yet another initiative for locals to experience the value of their biodiversity. Led by visionary scientists like Daniel Janzen (who used his own MacArthur grant money to purchase Costa Rican land to be set aside as a national park), local peoples have been trained, not only as guards and guides, but as parataxonomists, who sort and identify the myriad species of plants, insects, and other animals that populate their rain forests.

Such forms of employment dovetail nicely with an even better-known initiative, global ecotourism. Behind diamond mining and the cattle industry, ecotourism ranks third in the economy of Botswana, which, per capita, is one of the more well-off African nations. We have seen the collision between the cattle and wildlife in Botswana. Although it is perhaps still true that cattle (which are traditional to the Botswana culture, in addition to being a European-subsidized big business) hold the upper hand, it is not entirely lost on the people and the government of Botswana that tourism, especially tourism built around big-game viewing, is also bringing in sizable quantities of money into the country each year.

Ecotourism is no panacea. Scientists are well aware that no system can be observed without somehow the observer disturbing that system. This is all too true of ecotourism. With tourists come motor vehicles and planes, development and the spread of human waste. More subtly, but just as poignantly, the ecosystems themselves are disrupted directly. I will never forget being in an open vehicle on a game drive in the Okavango Delta when someone spotted two eagles sitting in a tree. The larger one was a huge, magnificent adult martial eagle. The other, smaller one was a tawny eagle. As we approached, the martial eagle flew off, clutching in its talons what was later identified as the remains of a civet cat. Off went the tawny in hot pursuit. Smaller birds are more maneuverable than larger ones, and this tawny was harassing the martial eagle, trying to get it to drop its meal.

The martial settled in a tree 45 meters away, but the tawny went for it, so the martial lumbered back toward the original tree where we were by then parked. Trying to land on a branch, it smacked its wing and tumbled to the ground, where for a moment, stunned, it huddled motionless. But then it was suddenly attacked, not this time by the tawny, but by a black-backed jackal whose den must have been in the grasses right near the tree. The eagle raised up, but its left wing, which seemed broken, jutted out at an odd angle. It couldn't fly. Injured, it had to stand its ground, turning to face the jackal as it circled the eagle, snapping at it. After a few minutes the jackal ran off, and the eagle just hunkered down.

We were stricken. It was obvious that our car had prompted the martial eagle to fly off, that we had triggered a flurry of exciting activity that had ended in the perhaps ultimately fatal injury of a truly magnificent top predator of the African wilds. The eagle just stayed there, and finally we backed away, headed back to camp for a typically luxurious lunch. We felt miserable.

The story has a happy ending. The camp manager immediately rode out to check on the eagle, prepared to bring it in, and even, if necessary, to get it to a facility where such injured birds are routinely cared for. It turned out not to be necessary. By the time she arrived, the eagle was perched back up in the tree, its wing folded normally. She waited until the eagle flew— again quite normally—and came back to tell us that all was right with this little corner of the world.

The moral remains. The upside of ecotourism is, I think, profound. Not only is money injected into the local economy (though how much of the money reaches truly local hands depends on the situation), but well-educated, intelligent travelers witness the way the world is. They may be there to see lions stalking antelope, but they cannot help but learn as well about the fragility of the ecosystems they are visiting and about the local manifestations of the global Sixth Extinction. But the downside is there as well. You simply cannot visit an ecosystem without disturbing it, as the martial eagle story so graphically illustrates. The more tourists, the greater the disruption. Costa Rica's Manuel Antonio Park, which has the highest density of visitors of the entire Costa Rican system of national parks, is beginning to fray—a victim, perhaps, of its success.

Perhaps the best model of how local economies can be turned—so that local ecosystems and species are viewed as assets rather than sources to be plundered or impediments to development—lies in the Campfire and similar movements, briefly encountered in chapter 1 on Botswana. The idea, however simple, is profound. Local peoples assume active ownership of the living resources around them. They can do absolutely what they want with them, with the understanding that overharvesting leads to loss of that very resource. They can promote ecotourism, allow hunting (for sport, for food), and utilize the system in any way they see fit. They have learned how to steward that land and those species. They know they must regulate all forms of usage, or that resource will disappear. Here is direct translation of sustainability and the need to meld conservation with the economic needs of people actually living there. Therein lies real hope for striking a balance.

6. We Must Develop a Political Will and Agenda. Conservation is not a new movement. Theodore Roosevelt was a conspicuous part of the effort to save the American bison from extinction and, perhaps more importantly, to create the federal national parks system. Though some historians argue that the U.S. national park system reflected the desire of railroad magnates to set aside scenic wonders as an incentive for railroad travel in the late nineteenth and early twentieth centuries, there is no doubt that the sheer act of setting aside vast tracts of acreage also greatly abetted conservationist goals. At mid-twentieth century, we have the writings of such inspirational conservationists as Aldo Leopold and Rachel Carson. The Earth Day Movement goes back to the 1960s, and there is no doubt that the air is basically cleaner around New York City, where I live, than it was several decades ago.

Nongovernmental organizations (NGOs) dedicated to lobbying and other pro-active conservation goals have sprung up, and, severally and collectively, have contributed much to the cause. Likewise, the academic scientific community has begun to take heed of the current crisis, and to respond. A new field, conservation biology, dedicated to the application of biological principles to the task of conserving species and ecosystems, has been developed. Entire institutes, such as the American Museum's Center for Biodiversity and Conservation, have been created with the avowed purpose of combating the Sixth Extinction through education (for example, graduate-level training in ecology and systematics of students from developing nations), field studies, and the development and promulgation of policies of action. Scientific societies are also getting involved—as in the consortium of plant and animal systematists, who have promulgated *Systematics Agenda 2000*, a well-conceived plan to inventory the world's biodiversity in a massive, integrated effort.

Most critically, the American public has, by degrees, become more sensitive to environmental issues and more concerned with the future of life and the physical environment around us. The back channels of cable television are crammed with nature shows, and if most of these fail even to mention the Sixth Extinction, they do signal an appreciation for the natural world. Though the United States lags behind in the formation of "Green" political parties typical of many different nations in Europe and elsewhere around the globe, environmental issues consistently rank high in national politics.

The United States—inexplicably and inexcusably, in my view—has yet to sign the Biodiversity Treaty that emanated in 1992 from the Rio Conference. Most of the world's nations have signed this document, which, among other things, calls for concerted study of each nation's biodiversity riches and a massive effort to slow the destruction of ecosystems and loss of the world's species. This failure is bipartisan, as Republican George Bush initially declined, and his successor, Democrat Bill Clinton, has so far also opted against signing.

What we need is bipartisan *positive* action. The good will of the American people, plus the efforts of individuals, nongovernmental organizations, NGOs, and scientific groups means a lot, but we have to put our money where our mouths are, and that means substantial and consistent governmental support. The old conservationist saw never meant so much as it now does: "Think globally, act locally." That remains true in its original intent. If everyone takes cares of his own neighborhood, the global environment will be the better off. Here, we move the action up a notch: The United States must assume a leadership role, taking care of its own critical environmental needs, its own loss of biodiversity, but also spearheading by example and direct action, a coordinated *global* movement aimed directly at staving off the Sixth Extinction. As one national politician recently put it: Assuming we start with 8 million species of insects, we have to ask ourselves, do we want a future with 2 million species or with, say, 6 million? It is too late to have all 8 million, but clearly 6 million is preferable to 2 million.

This means thinking about the future, several generations down the pike, rather than tomorrow's short-term gains. Politicians, like the rest of us, often have difficulty thinking beyond the immediate. The effort that is needed—the funding and the leadership—involves (1) implementation of suggested initiatives to document existing/remaining biodiversity globally; (2) assessment of the extent and health of remaining ecosystems; (3) simultaneous development of conservation policies that (a) reflect the true economic value of the living world and (b) reflect as well the economic needs of existing human populations interacting with these ecosystems (see above, agenda point 5); and (4) implementation of these policies. The task is enormous, will cost billions of dollars, and must involve the direct participation not only of politicians and life scientists, but businessmen, economists, demographers, lawyers, sociologists, and anthropologists—any and all with expertise on the relation of humans to the physical and living environment.

People ask me, What's going to happen? I, of course, don't really know—and I confess to some days of pure pessimism cutting into a generally optimistic outlook. I do know that we have arrived at this state—worrisome as it is—simply by doing what we do best: using our brains, our culture, to wrest a living from the natural world. Ecologists tend to measure success by population size. By those standards, we are becoming more successful all the time. We know that we are eating ourselves out of house and home and despoiling our nest in the bargain. We must remember that we got to this predicament honestly, by devising bigger and better "mousetraps"—contrivances to make a living. The biggest one so far, the one that utterly changed our position in the natural world, was the invention of agriculture, an experiment that, if it has perhaps succeeded too well, nonetheless has given us all the finer things of human experience.

And so I think: If we have arrived at this exalted/troubled state through our own cleverness, surely we are smart enough to call a halt, to say enough, to stabilize, *to strike a balance.*

Appendix I ANIMAL SPECIES EXTINCT SINCE CIRCA 1600

SPECIES	ENGLISH NAME	DISTRIBUTION	LAST RECORDED	POSSIBLE CAUSE
CORALS (CNIDARIA)				
Order MILLEPORINA				
Family Milleporidae				
Millepora sp.		Panama	1983	
MOLLUSKS				
Order ARCHAEOGASTROPODA				
Family Acmaeidae				
Lottia alveus	Eelgrass Limpet	USA		B
Order MESOGASTROPODA				
Family Hydrobiidae				
Bythiospeum pfeifferi		Austria		
Clappia umbilicata	Umbilicate Pebblesnail	USA		
Ohridohauffenia drimica		Yugoslavia	1980s	
Family Pleuroceridae				
Elimia clausa	Closed Elimia	USA		B
Elimia fusiformis	Fusiform Elimia	USA		B
Elimia hartmaniana	High-spired Elimia	USA		B
Elimia impressa	Constricted Elimia	USA		B
Elimia jonesi	Hearty Elimia	USA		B
Elimia laeta	Ribbed Elimia	USA		B
Elimia pilsbryi	Rough-lined Elimia	USA		B
Elimia pupaeformis	Pupa Elimia	USA		B
Elimia pygmaea	Pygmy Elimia	USA		B
Elimia varians	Puzzle Elimia	USA		B
Gyrotoma incisa	Excised Slitshell	USA	1924	
Gyrotoma lewisii	Striate Slitshell	USA	1924	
Gyrotoma pagoda	Pagoda Slitshell	USA	1924	
Gyrotoma pumila	Ribbed Slitshell	USA	1924	
Gyrotoma pyramidata	Pyramid Slitshell	USA	1924	
Gyrotoma walkeri	Round Slitshell	USA	1924	
Leptoxis clipeata	Agate Rocksnail	USA		
Leptoxis formanii	Interrupted Rocksnail	USA		
Leptoxis ligata	Rotund Rocksnail	USA		
Leptoxis lirata	Lirate Rocksnail	USA		
Leptoxis occultata	Bigmouth Rocksnail	USA		
Leptoxis showalterii	Coosa Rocksnail	USA		
Leptoxis vittata	Striped Rocksnail	USA		
Family Pomatiasidae				
Tropidophora carinata		Mauritius	1881	B
Order STYLOMMATOPHORA				
Family Endodontidae				
Discus guerinianus		Madeira (Portugal)	1870s	
Kondoconcha othnius		Rapa (F. Polynesia)	1934	
Libera subcavernula		Raratonga (Cook Is)	1880s	
Libera tumuloides		Raratonga (Cook Is)	1880s	
Mautodonta acuticosta		Raiatea (F. Polynesia)	1880s	
Mautodonta boraborensis		Borabora (F. Polynesia)	1880s	
Mautodonta ceuthma		Raivavae (F. Polynesia)	1880s	
Mautodonta consimilis		Raiatea (F. Polynesia)	1880s	
Mautodonta consobrina		Huahine (F. Polynesia)	1880s	
Mautodonta maupiensis		Maupiti (F. Polynesia)	1880s	

SPECIES	ENGLISH NAME	DISTRIBUTION	LAST RECORDED	POSSIBLE CAUSE
		MOLLUSKS (Continued)		
Mautodonta parvidens		Society Is (F. Polynesia)	1880s	
Mautodonta punctiperforata		Moorea (F. Polynesia)	1880s	
Mautodonta saintjohni		Borabora (F. Polynesia)	1880s	
Mautodonta subtilis		Huahine (F. Polynesia)	1880s	
Mautodonta unilamellata		Raratonga (Cook Is)	1880s	
Mautodonta zebrina		Raratonga (Cook Is)	1880s	
• *Opanara altiapica*		Rapa (F. Polynesia)	1934	
• *Opanara areaensis*		Rapa (F. Polynesia)	1934	
• *Opanara bitridentata*		Rapa (F. Polynesia)	1934	
• *Opanara caliculata*		Rapa (F. Polynesia)	1934	
• *Opanara depasoapicata*		Rapa (F. Polynesia)	1934	
• *Opanara duplicidentata*		Rapa (F. Polynesia)	1934	
• *Opanara fosbergi*		Rapa (F. Polynesia)	1934	
• *Opanara megomphala*		Rapa (F. Polynesia)	1934	
• *Opanara perahuensis*		Rapa (F. Polynesia)	1934	
• *Orangia cooki*		Rapa (F. Polynesia)	1934	
• *Orangia maituatensis*		Rapa (F. Polynesia)	1934	
• *Orangia sporadica*		Rapa (F. Polynesia)	1934	
* *Pilula cycloria*		Mauritius		
• *Rhysoconcha atanuiensis*		Rapa (F. Polynesia)	1934	
• *Rhysoconcha variumbilicata*		Rapa (F. Polynesia)	1934	
• *Ruatara koarana*		Rapa (F. Polynesia)	1934	
• *Ruatara oparica*		Rapa (F. Polynesia)	1934	
Taipidon anceyana		Hiva Oa (F. Polynesia)	1880s	
Taipidon marquesana		Nuku Hiva (F. Polynesia)	1880s	
Taipidon octolamellata		Hiva Oa (F. Polynesia)	1880s	
Thaumatodon multilamellatus		Raratonga (Cook Is)	1880s	
Family Bulimulidae				
Amphibulima patula		Guadeloupe		B
Bulimulus duncanus		Galapagos (Ecuador)	late 1800s	H
Leuchocharis loyaltyensis		New Caledonia	1900s	B
Leuchocharis porphyrocheila		New Caledonia	1900s	B
Family Charopidae				
Helenoconcha leptalea		St Helena	1870s	
Helenoconcha minutissima		St Helena	1870s	
Helenoconcha polyodon		St Helena	1870s	
Helenoconcha pseustes		St Helena	1870s	
Helenoconcha sexdentata		St Helena	1870s	
Helenodiscus bilamellata		St Helena	1870s	
Helenodiscus vernoni		St Helena	1870s	
Pseudohelenoconcha dianae		St Helena	1870s	
Pseudohelenoconcha laetissima		St Helena	1870s	
Pseudohelenoconcha persoluta		St Helena	1870s	
Pseudohelenoconcha spurca		St Helena	1870s	
Sinployea canalis		Raratonga (Cook Is)	1872	
Sinployea decorticata		Raratonga (Cook Is)	1872	
Sinployea harveyensis		Raratonga (Cook Is)	1872	
Sinployea otareae		Raratonga (Cook Is)	1872	
Sinployea planospira		Raratonga (Cook Is)	1872	
Sinployea proxima		Raratonga (Cook Is)	1872	
Sinployea rudis		Raratonga (Cook Is)	1872	
Sinployea tenuicostata		Raratonga (Cook Is)	1872	
Sinployea youngi		Raratonga (Cook Is)	1872	
Family Achatinellidae				
Achatinella abbreviata		Hawaii (USA)	1963	A,C,D
Achatinella buddii		Hawaii (USA)	early 1900s	A,C,D
Achatinella caesia		Hawaii (USA)	early 1900s	A,C,D
Achatinella casta		Hawaii (USA)		A,C,D

SPECIES	ENGLISH NAME	DISTRIBUTION	LAST RECORDED	POSSIBLE CAUSE
MOLLUSKS (Continued)				
Achatinella decora		Hawaii (USA)	early 1900s	A,C,D
Achatinella elegans		Hawaii (USA)	1952	A,C,D
Achatinella juddii		Hawaii (USA)	1958	A,C,D
Achatinella juncea		Hawaii (USA)		A,C,D
Achatinella lehuiensis		Hawaii (USA)	1922	A,C,D
Achatinella papyracea		Hawaii (USA)	1945	A,C,D
Achatinella rosea		Hawaii (USA)	1949	A,C,D
Achatinella spaldingi		Hawaii (USA)	1938	A,C,D
Achatinella stewarti		Hawaii (USA)	1961	A,C,D
Achatinella thaanumi		Hawaii (USA)	1900s	A,C,D
Achatinella valida		Hawaii (USA)	1951	A,C,D
Achatinella vittata		Hawaii (USA)	1953	A,C,D
x *Elasmias jauffreti*		Rodrigues (Mauritius)		
x *Elasmias* sp.		Mauritius		
Partulina crassa		Hawaii (USA)	1914	C?
Partulina montagui		Hawaii (USA)	1913	C?
Family Partulidae				
Partula exigua	Moorean Viviparous Tree Snail	Moorea (F. Polynesia)	1977	C?
Partula filosa	Tahiti Viviparous Tree Snail	Tahiti (F. Polynesia)		
Partula producta	Tahiti Viviparous Tree Snail	Tahiti (F. Polynesia)		
Partula salifana		Guam		
Samoana abbreviata		American Samoa	1940	B
Family Amastridae				
Carelia anceophila		Hawaii (USA)	1930	B,C,D
Carelia bicolor		Hawaii (USA)	1970	B,C,D
Carelia cumingiana		Hawaii (USA)	1930	B,C,D
Carelia glossema		Hawaii (USA)	1930	B,C,D
Carelia kalalauensis		Hawaii (USA)	1945/47	B,C,D
Carelia knudseni		Hawaii (USA)	1930	B,C,D
Carelia olivacea		Hawaii (USA)	1930	B,C,D
Carelia paradoxa		Hawaii (USA)	1930	B,C,D
Carelia periscelis		Hawaii (USA)	1930	B,C,D
Carelia tenebrosa		Hawaii (USA)	1930	B,C,D
Carelia turricula		Hawaii (USA)	1930	B,C,D
Family Vertiginidae				
Campolaemus perexilis		St Helena	1870s	
Nesopupa turtoni		St Helena	1870s	
Family Pupillidae				
x *Gibbulinopsis* sp.		Rodrigues (Mauritius)		
Leiostyla abbreviata		Madeira (Portugal)	1870s	
Leiostyla cassida		Madeira (Portugal)	1870s	
Leiostyla concinna		Madeira (Portugal)	1870s	
Leiostyla gibba		Madeira (Portugal)	1870s	
Leiostyla laevigata		Madeira (Portugal)	1870s	
Leiostyla lamellosa		Madeira (Portugal)	1870s	
Leiostyla simulator		Madeira (Portugal)	1870s	
Pupa obliquicostata		St Helena	1870s	
Family Helixarionidae				
Colparion madgei		Rodrigues (Mauritius)	1938	B
Ctenoglypta newtoni		Mauritius	1871	B
x *Ctenophila planorbina*		Mauritius		
Diastole matafaoi		American Samoa	1940	D?
x *Erepta thiriouxi*		Mauritius		
x *Erepta* sp.		Mauritius		
Pachystyla ruforonata		Mauritius	1869	B
x *Plegma bewsheri*		Rodrigues (Mauritius)		
x *Plegma duponti*		Mauritius		
x *Plegma* sp.		Mauritius		

SPECIES	ENGLISH NAME	DISTRIBUTION	LAST RECORDED	POSSIBLE CAUSE
MOLLUSKS (Continued)				
Family Ferussaciidae				
Cecilioides eulima		Madeira (Portugal)	1870s	
Family Subulinidae				
Chilonopsis blofeldi		St Helena	1870s	
Chilonopsis exulatus		St Helena	1870s	
Chilonopsis helena		St Helena	1870s	
Chilonopsis melanoides		St Helena	1870s	
Chilonopsis nonpareil		St Helena	1870s	
Chilonopsis subplicatus		St Helena	1870s	
Chilonopsis subtruncatus		St Helena	1870s	
Chilonopsis turtoni		St Helena	1870s	
Family Helicidae				
Discula lyelliana		Madeira (Portugal)	1870s	
Discula tetrica		Madeira (Portugal)	1870s	
Geomitra delphinuloides		Madeira (Portugal)	1870s	
Lemniscia galeata		Madeira (Portugal)	1870s	
Pseudocampylaea lowei		Madeira (Portugal)	late 19th C.	
Family Streptaxidae				
Edentulina thomasetti		Seychelles	1908	
Gibbus lyonetianus		Mauritius	1905	B
Gonidomus newtoni		Mauritius	1867	B
x *Gonospira cirneensis*		Mauritius		
x *Gonospira heliodes*		Mauritius		
x *Gonospira majusculus*		Mauritius		
Imperturbata violescens?		Seychelles		
Family Assimineidae				
x *Omphalotropis plicosa*		Mauritius	1878	B
x *Omphalotropis caldwelli*		Mauritius		
x *Omphalotropis dupontiana*		Mauritius		
x *Omphalotropis maxima*		Mauritius		
x *Omphalotropis multilirata*		Mauritius		
x *Omphalotropis* sp.		Mauritius		
Family Pomatiasidae				
x *Tropidophora bewsheri*		Rodrigues (Mauritius)		
x *Tropidophora bipartita*		Rodrigues (Mauritius)		
x *Tropidophora deflorata*		Réunion		
x *Tropidophora lienardi*		Mauritius		
x *Tropidophora mauritiana*		Mauritius		
Order UNIONOIDA				
Family Unionidae				
Alasmidonta mccordi	Coosa Elktoe	USA		
Alasmidonta wrightiana	Ochlacknee Arc-mussel	USA		
Epioblasma arcaeformis	Sugarspoon	USA	1940s	B
Epioblasma biemarginata	Angled Riffleshell	USA	1960s	B
Epioblasma flexuosa	Leafshell	USA	1940s	B
Epioblasma haysiana	Acornshell	USA		
Epioblasma lenior	Narrow Catspaw	USA	1965	B
Epioblasma lewisii	Forkshell	USA	1964	B
Epioblasma personata	Round Combshell	USA	1930	B
Epioblasma propinqua	Tennessee Riffleshell	USA	1930	B
Epioblasma sampsonii	Wabask Riffleshell	USA	1950s/60s	B
Epioblasma stewardsoni	Cumberland Leafshell	USA	1930	B
Medionidus mcglameriae	Tombigbee Moccasinshell	USA		
CRUSTACEANS				
Order AMPHIPODA				
Family Crangonyctidae				
Stygobromus hayi	Hay's Spring Scud	USA	1957	
Stygobromus lucifugus	Rubious Cave Amphipod	USA		

SPECIES	ENGLISH NAME	DISTRIBUTION	LAST RECORDED	POSSIBLE CAUSE
CRUSTACEANS (Continued)				
Order DECAPODA				
Family Astacidae				
Pacifastacus nigrescens	Sooty Crayfish	USA	1860s	
Family Atyidae				
Syncaris pasadenas	Pasadena Freshwater Shrimp	USA	1933	
INSECTS				
Order EPHEMEROPTERA				
Family Siphlonuridae				
Acanthometropus pecatonia	Pecatonica River Mayfly	USA	1927	
Family Ephemeridae				
Pantagenia robusta	Robust Burrowing Mayfly	USA		
Order ORTHOPTERA				
Family Tettigoniidae				
Neduba extincta	Antioch Dunes Shieldback Katydid	USA	1937	
Order PHASMATOPTERA				
Family Phasmatidae				
Dryococelus australis	Lord Howe Island Stick-insect	Lord Howe I (Australia)	1969	
Order DERMAPTERA				
Family Labiduridae				
* *Labidura herculeana*	St Helena Earwig	St Helena	1967	
Order PLECOPTERA				
Family Chloroperlidae				
Alloperla roberti	Robert's Stonefly	USA		
Order HOMOPTERA				
Family Pseudococcidae				
Clavicoccus erinaceus		Hawaii (USA)		
Phyllococcus oahuensis		Hawaii (USA)		
Order COLEOPTERA				
Family Cerambycidae				
Xyloteles costatus	Pitt Island Longhorn Borer	Chatham I (NZ)	1930s	B,C
Family Curculionidae				
Dryophthorus distinguendus		Hawaii (USA)		
Dryotribus mimeticus		Hawaii (USA)		
Hadramphus tuberculatus		New Zealand	1910	C
Macrancylus linearis		Hawaii (USA)		
Oedemasylus laysanensis		Hawaii (USA)		
Pentarthrum blackburnii		Hawaii (USA)		
Rhyncogonus bryani		Hawaii (USA)		
Family Carabidae				
* *Aplothorax burchelli*		St Helena	1967?	
* *Mecodema punctellum*		Stephens I (NZ)		B,G
Order DIPTERA				
Family Tabanidae				
Stonemyia volutina	Volutine Stoneyian Tabanid Fly	USA		
Family Dolichopodidae				
Campsicnemus mirabilis		Hawaii (USA)		
Family Drosophilidae				
Drosophila lanaiensis		Hawaii (USA)		
Order TRICHOPTERA				
Family Rhyacophilidae				
Rhyacophila amabilis	Castle Lake Caddis-fly	USA		
Family Hydropsychidae				
Hydropsyche tobiasi	Tobias' Caddis-fly	Germany	1920s	
Family Leptoceridae				
Triaenodes phalacris	Athens Caddis-fly	USA		

SPECIES	ENGLISH NAME	DISTRIBUTION	LAST RECORDED	POSSIBLE CAUSE
INSECTS (Continued)				
Triaenodes tridonata	Three-tooth Caddis-fly	USA		
Order LEPIDOPTERA				
Family Zygaenidae				
Levuana iridescens	Levuana Moth	Fiji	1929	E
Family Lycaenidae				
Glaucopsyche xerces	Xerces Blue	USA	early 1940s	
Family Libytheidae				
Libythea cinyras		Mauritius	1865	
Family Nymphalidae				
Euthalia malapana		Taiwan		
Family Pyralidae				
Genophantis leahi		Hawaii (USA)	early 1900s	
Hedylepta asaphombra		Hawaii (USA)	1970s	
Hedylepta coninuatalis		Hawaii (USA)	1958	
Hedylepta epicentra		Hawaii (USA)	early 1900s	
+ *Hedylepta euryprora*		Hawaii (USA)		E
+ *Hedylepta fullawayi*		Hawaii (USA)		E
Hedylepta laysanensis		Hawaii (USA)		
+ *Hedylepta meyricki*		Hawaii (USA)		E
+ *Hedylepta musicola*		Hawaii (USA)		E
Hedylepta telegrapha		Hawaii (USA)		
Oeobia sp.		Hawaii (USA)	1911	
Family Geometridae				
Scotorhythra nesiotes		Hawaii (USA)	early 1900s	
Scotorhythra megalophylla		Hawaii (USA)	early 1900s	
Scotorhythra paratactis		Hawaii (USA)	early 1900s	
Tritocleis microphylla		Hawaii (USA)	1890s	
Family Sphingidae				
Manduca blackburni		Hawaii (USA)	1960s	
Family Noctuidae				
Agrotis crinigera	Poco Noctuid Moth	Hawaii (USA)	1926	E
Agrotis fasciata	Midway Noctuid Moth	Hawaii (USA)		
Agrotis kerri		Hawaii (USA)	1923	
Agrotis laysanensis		Hawaii (USA)	1911	
Agrotis photophila		Hawaii (USA)		
Agrotis procellaris		Hawaii (USA)	pre-1900	
Helicoverpa confusa		Hawaii (USA)	post-1927	
Helicoverpa minuta	Minute Noctuid Moth	Hawaii (USA)	pre-1911	
Hypena laysanensis	Laysan Dropseed Noctuid Moth	Hawaii (USA)	1911	
+ *Hypena newelli*		Hawaii (USA)		
+ *Hypena plagiota*		Hawaii (USA)		
+ *Hypena senicula*		Hawaii (USA)		
Peridroma porphyrea		Hawaii (USA)		
Order HYMENOPTERA				
Family Colletidae				
Nesoprosopis angustula	Lanai Yellow-faced Bee	Hawaii (USA)		
Nesoprosopis blackburni	Blackburn's Yellow-faced Bee	Hawaii (USA)		
Nesoprosopis connectens	Connected Yellow-faced Bee	Hawaii (USA)		
FISHES				
Order PETROMYZONTIFORMES				
Family Petromyzontidae				
Lampetra minima	Miller Lake Lamprey	USA	1953	E
Order CYPRINIFORMES				
Family Cyprinidae				
Evarra bustamantei		Mexico	1970	B
Evarra eigenmanni		Mexico	1970	B
Evarra tlahuacensis		Mexico	1970	B

SPECIES	ENGLISH NAME	DISTRIBUTION	LAST RECORDED	POSSIBLE CAUSE
FISHES (Continued)				
Gila crassicauda	Thicktail Chub	USA	1957	B,C/D
Lepidomeda altivelis	Pahranagat Spinedace	USA	1940	C/D
Notropis amecae	Ameca Shiner	Mexico	1970	C/D
Notropis aulidion	Durango Shiner	Mexico	1965	C/D
Notropis orca	Phantom Shiner	Mexico, USA	1975	B,C/D
Pogonichthys ciscoides	Clear Lake Splittail	USA	1970	B,C/D
Rhinichthys deaconi	Las Vegas Dace	USA	1955	B
Stypodon signifer	Stumptooth Minnow	Mexico	1930	B
Family Catostomidae				
Chasmistes muriei	Snake River Sucker	USA	1928	B
Lagochila lacera	Harelip Sucker	USA	1910	G
Order SALMONIFORMES				
Family Retropinnidae				
* *Prototroctes oxyrhynchus*	New Zealand Grayling	New Zealand	1920s	B,D,H
Family Salmonidae				
Coregonus alpenae	Longjaw Cisco	USA, Canada	1978	A,C
Coregonus johannae	Deepwater Cisco	USA, Canada	1955	A,C/D
Salvelinus agassizi	Silver Trout	USA	1930	A,C/D
Order CYPRINODONTIFORMES				
Family Fundulidae				
Fundulus albolineatus	Whiteline Topminnow	USA	1900	B,C/D
Family Poeciliidae				
Gambusia amistadensis	Amistad Gambusia	USA	1973	B
* *Gambusia georgei*	San Marcos Gambusia	USA	1983	B,C/D
* *Priapella bonita*	Guayacon Ojiazul	Mexico		
Family Goodeidae				
Characodon garmani	Parras Characodon	Mexico	1900	B?
Empetrichthys merriami	Ash Meadows Killifish	USA	1953	B,C/D
Family Cyprinodontidae				
Cyprinodon latifasciatus	Perrito de Parras	Mexico	1930	B
Cyprinodon sp.	Monkey Spring Pupfish	USA	1971	B,C/D
Cyprinodon sp.		Mexico		
Cyprinodon sp.		Mexico		
Order SCORPAENIFORMES				
Family Cottidae				
Cottus echinatus	Utah Lake Sculpin	USA	1928	B,C/D
AMPHIBIANS				
Order ANURA				
Family Discoglossidae				
Discoglossus nigriventer	Israel Painted Frog	Israel	1940	B
Rana fisheri	Relict Leopard Frog	USA	1960	B
REPTILES				
Order TESTUDINES				
Family Testudinidae				
Cylindraspis borbonica		Réunion	1800	
Cylindraspis indica		Réunion	1800	A
Cylindraspis inepta		Mauritius	early 18th C.	A,C/D
Cylindraspis peltastes		Rodrigues (Mauritius)	1800	A,B,C/D
Cylindraspis triserrata		Mauritius	early 18th C.	A,C/D
Cylindraspis vosmaeri		Rodrigues (Mauritius)	1800	A,C/D
Order SAURIA				
Family Gekkonidae				
Hoplodactylus delcourti		New Zealand (?)	mid 19th C.?	
Phelsuma edwardnewtoni	Newton's Day Gecko	Rodrigues (Mauritius)	1917	C
Phelsuma gigas	Giant Day Gecko	Rodrigues (Mauritius)	end 19th C.	C
Family Iguanidae				
Leiocephalus eremitus		Navassa I (USA)	1900	C
Leiocephalus herminieri		Martinique	1830s	

SPECIES	ENGLISH NAME	DISTRIBUTION	LAST RECORDED	POSSIBLE CAUSE
REPTILES (Continued)				
Family Teiidae				
Ameiva cineracea		Guadeloupe	early 20th C.	
* *Ameiva major*	Martinique Giant Ameiva	Martinique		C?
Family Anguidae				
Celestus occiduus	Jamaican Giant Galliwasp	Jamaica	1840	C?
Family Scincidae				
# *Leiolopisma mauritiana*		Mauritius	1600	C
Macroscincus coctei	Cape Verde Giant Skink	Cape Verde	early 20th C.	A?
* *Tiliqua adelaidensis*	Adelaide Pigmy Bluetongue	Australia	1959	B,C
Order SERPENTES				
Family Boidae				
* *Bolyeria multocarinata*		Round I (Mauritius)	1975	
Family Typhlopidae				
Typhlops cariei		Mauritius	17th C.	C
Family Colubridae				
* *Alsophis ater*	Jamaican Tree Snake	Jamaica	1950	A,C
Alsophis sancticrucis	St Croix Racer	Virgin Is (US)	20th C.	A,C
* *Liophis cursor*	Martinique Racer	Martinique	1963	C
* *Liophis perfuscus*	Barbados Racer	Barbados	mid 20th C.?	C
BIRDS				
Order STRUTHIONIFORMES				
Family Dromaiidae				
Dromaius diemenianus	Kangaroo Island Emu	Kangaroo I (Australia)	1803	B
Family Aepyornithidae				
Aepyornis maximus	Great Elephantbird	Madagascar	1650	A,B
Family Anomalopterygidae				
Dinornis torosus	Brawny Great Moa	New Zealand	1670	A,B
Eurapteryx gravis	Burly Lesser Moa	New Zealand	1640	A,B
Megalaperyx didinus	South Island Tokoweka	New Zealand	1785	A,B
Order GALLIFORMES				
Family Phasianidae				
Coturnix novaezelandiae	New Zealand Quail	New Zealand	1875	F
Ophrysia superciliosa	Himalayan Mountain Quail	India	1868	A
Order ANSERIFORMES				
Family Anatidae				
Alopochen mauritianus	Mauritian Shelduck	Mauritius	1698	
Anas theodori	Mauritian Duck	Mauritius, ?Réunion	1696	
Camptorhynchus labradorius	Labrador Duck	Canada, USA	1878	A,B
Cygnus sumnerensis	Chatham Island Swan	Chatham I (NZ)	1590-1690	
Mergus australis	Auckland Island Merganser	New Zealand	1905	A,B,C
* *Rhodonessa caryophyllace*a	Pink-headed Duck	India, Nepal	1935	A
Sheldgoose sp.		Réunion	1674	
Order CORACIIFORMES				
Family Alcedinidae				
Halcyon miyakoensis	Ryukyu Kingfisher	Nansei-shoto (Japan)	1841	
Order CUCULIFORMES				
Family Cuculidae				
* *Coua delalandei*	Snail-eating Coua	Madagascar	1930	A,B,C/D
Order PSITTACIFORMES				
Family Psittacidae				
Anodorhynchus glaucus	Glaucous Macaw	Brazil, Uruguay	1955	
Ara tricolor	Cuban Red Macaw	Cuba	1885	A,E
Charmosyna diadema	New Caledonia Lorikeet	New Caledonia	1860	B
Conuropsis carolinensis	Carolina Parakeet	USA	1914	E
Cyanoramphus ulietanus	Raiatea Parakeet	Raiatea (F. Polynesia)	1773	
Cyanoramphus zealandicus	Black-fronted Parakeet	Tahiti (F. Polynesia)	1844	B
'Lophopsittacus' bensoni	Mauritius Grey Parrot	Mauritius	1765	C/D
Lophopsittacus mauritianus	Mauritius Parrot	Mauritius	1675	A,C

SPECIES	ENGLISH NAME	DISTRIBUTION	LAST RECORDED	POSSIBLE CAUSE
		BIRDS (Continued)		
Mascarinus mascarinus	Mascarene Parrot	Réunion	1775 (1834 in captivity)	B
'Necropsittacus' rodericanus	Rodrigues Parrot	Rodrigues (Mauritius)	1761	A,C/D
Nestor productus	Norfolk Island Kaka	Phillip I (Australia)	1851	A,E
Psittacula exsul	Rodrigues Ring-necked Parakeet	Rodrigues (Mauritius)	1876	B
Psittacula wardi	Seychelles Alexandrine Parrot	Seychelles	1870	A,B
Order TROCHILIFORMES				
Family Trochilidae				
Chlorostilbon bracei	New Providence Hummingbird	Bahamas	1877	
Family Caprimulgidae				
* *Siphonorhis americanus*	Jamaica Least Pauraque	Jamaica	1859	C
Order STRIGIFORMES				
Family Strigidae				
Athene blewitti	Forest Owlet	India	1914	
'Athene' murivora	Rodrigues Little Owl	Rodrigues (Mauritius)	1726	B
?Sauzieri sp.	Mauritian Owl	Mauritius		
* *Sceloglaux albifacies*	Laughing Owl	New Zealand	1914	B,C
'Scops' commersoni	Mauritian Owl	Mauritius	1836	
Family Aegothelidae				
Aegotheles savesi	New Caledonia Owlet-frogmouth	New Caledonia	1880	
Order COLUMBIFORMES				
Family Raphidae				
'Ornithaptera' solitaria	Réunion Solitaire	Réunion	1710-1715	A
Pezophaps solitarius	Rodrigues Solitaire	Rodrigues (Mauritius)	1765	A
Raphus cucullatus	Dodo	Mauritius	1665	A,C,D
Family Columbidae				
Alextroenas nitidissima	Pigeon Hollandais	Mauritius	1835	A,C
'Alextroenas' rodericana	Rodrigues Pigeon	Rodrigues (Mauritius)	1726	C/D
Columba jouyi	Ryukyu Wood Pigeon	Nansei-shoto (Japan)	1936	B
Columba versicolor	Bonin Wood Pigeon	Ogasawara-shoto (Japan)	1889	C
Ectopistes migratorius	Passenger Pigeon	USA	1914	A,B
* *Microgoura meeki*	Solomon I Crowned-pigeon	Choiseul (Solomon I)	1904	C
* *Ptilinopus mercierii*	Marquesas Fruit-dove	Marquesas I (F. Polynesia)	1922	C/D
Order GRUIFORMES				
Family Rallidae				
Aphanapteryx bonasia	Red Rail	Mauritius	1700	A,C/D
Aphanapteryx leguati	Rodrigues Rail	Rodrigues (Mauritius)	1761	
Atlantisia elpenor	Ascension Flightless Crake	Ascension I (UK)	1656	G(A,C)
Fulica newtoni	Mascarene Coot	Mauritius, Réunion	1693	
Gallinula nesiotis	Tristan Moorhen	Tristan da Cunha (UK)	1875-1900	C
Gallinula pacifica	Samoan Woodhen	Savaii (W. Samoa)	1908-1926	C
Gallirallus pacificus	Tahiti Rail	French Polynesia	1773/4	
Nesoclopeus woodfordi	Woodford's Rail	Bougainville (Papua New Guinea)	1936	
Porphyrio albus	Lord Howe Purple Gallinule	Lord Howe I (Australia)	1834	A
Porzana monasa	Kosrae Crake	Federated States of Micronesia	1827	C
Porzana palmeri	Laysan Rail	Hawaii (USA)	1944	C,D
Porzana sandwichensis	Hawaiian Rail	Hawaii (USA)	1898	C
Rallus dieffenbachii	Chatham Island Banded Rail	Chatham I (NZ)	1840	B,C
Rallus modestus	Chatham Island Rail	Chatham I (NZ)	1900	D
Rallus wakensis	Wake Island Rail	Wake I (USA)	1945	A
* *Tricholimnas lafresnayanus*	New Caledonia Rail	New Caledonia	1904	

SPECIES	ENGLISH NAME	DISTRIBUTION	LAST RECORDED	POSSIBLE CAUSE

BIRDS (Continued)

Order CICONIIFORMES

Family Scolopacidae

Prosobonia leucoptera	Tahitian Sandpiper	Tahiti, Moorea (F. Polynesia)	1773	D

Family Charadriidae

Haematopus meadewaldoi	Canarian Black Oystercatcher	Canary Is (Spain)	1913	G
Vanellus macropterus	Javanese Wattled Lapwing	Java (Indonesia)	1940	A,B

Family Laridae

Alca impennis	Great Auk	Canada, Iceland, Faeroes UK, Russia, Greenland	1844	A

Family Falconidae

Falco sp.		Réunion	1674	
Polyborus lutosus	Guadalupe Caracara	Guadalupe (Mexico)	1900	A,D,E

Family Podicipedidae

Podiceps andinus	Colombian Grebe	Colombia	1977	
Podilymbus gigas	Atitlan Grebe	Guatemala	1980-1986/7	A,D
Tachybaptus rufolarvatus	Lake Alaotra Grebe	Madagascar		

Family Phalacrocoracidae

Phalacrocorax perspicillatus	Spectacled Cormorant	Bering Straits (Russia)	1852	A

Family Ardeidae

Ixobrychus novaezelandia	New Zealand Little Bittern	New Zealand	1900	
Nycticorax mauritianus	Mauritius Night-heron	Mauritius	by 1700	
Nycticorax megacephalus	Rodrigues Night-heron	Rodrigues (Mauritius)	1761	
Nycticorax sp.		Réunion	by 1700	

Family Threskiornithidae

Borbonibis latipes	Réunion Flightless Ibis	Réunion	1773	

Family Ciconiidae

Ciconia sp.		Réunion	1674	

Family Procellariidae

* *Oceanodroma macrodactyla*	Guadalupe Storm-petrel	Guadalupe (Mexico)	1912-1922	C
Pterodroma sp.		Rodrigues (Mauritius)	1726	

Order PASSERIFORMES

Family Acanthisittidae

Xenicus longipes	Bush Wren	New Zealand	1972	B,C
Xenicus lyalli	Stephens Island Wren	Stephens I (NZ)	1874	C

Family Pycnonotidae

Hypsipetes sp.		Rodrigues (Mauritius)	1600s?	

Family Muscicapidae

Acrocephalus familiaris	Laysan Millerbird	Hawaii (USA)	1912-1923	B,D
Eutrichomyias rowleyi	Caerulean Paradise-flycatch	Sangihe (Indonesia)	1978	B
Myiagra freycineti	Guam Broadbill	Guam	1983	
* *Turnagra capensis*	Piopio	New Zealand	1955	B,C
Turdus ravidus	Grand Cayman Thrush	Cayman Is	1938	B
Zoothera terrestris	Kittlitz's Thrush	Ogasawara-shoto (Japan)	1928	C
Babbler sp.		Rodrigues (Mauritius)	1600s?	

Family Dicaeidae

Dicaeum quadricolor	Four-coloured Flowerpecker	Cebu (Philippines)	1906	B

Family Zosteropidae

Zosterops strenua	Lord Howe White-eye	Lord Howe I (Australia)	1928	A,B,C/D

Family Meliphagidae

Chaetoptila angustipluma	Kioea	Hawaii (USA)	1860	B
Moho apicalis	Oahu Oo	Hawaii (USA)	1837	A,B,C/D
* *Moho nobilis*	Hawaii Oo	Hawaii (USA)	1934	A,B,C/D
Ciridops anna	Ula-ai-hawane	Hawaii (USA)	1892	
Drepanis funerea	Black Mamo	Hawaii (USA)	1907	
Drepanis pacifica	Hawaii Mamo	Hawaii (USA)	1899	A,B
* *Hemignathus obscurus*	Akialoa	Hawaii (USA)	1960	

SPECIES	ENGLISH NAME	DISTRIBUTION	LAST RECORDED	POSSIBLE CAUSE
BIRDS (Continued)				
Hemignathus sagittirostris	Greater Amakihi	Hawaii (USA)	1900	B
* *Paroreomyza flammea*	Kakawihie Creeper	Hawaii (USA)	1963	
Psittirostra kona	Kona Grosbeak	Hawaii (USA)	1894	
Rhodacanthis flaviceps	Lesser Koa-finch	Hawaii (USA)	1891	
Rhodacanthis palmeri	Greater Koa-finch	Hawaii (USA)	1896	
Family Icteridae				
Quiscalus palustris	Slender-billed Grackle	Mexico	1910	B
Family Ploceidae				
Foudra sp.	Reunion Fody	Réunion	1671	
Family Fringillidae				
Chaunoproctus ferreorostris	Bonin Grosbeak	Ogasawara-shoto (Japan)	1890	B,C/D
Spiza townsendi	Townsend's Finch	USA	1833	
Family Sturnidae				
Aplonis corvina	Kosrae Mountain Starling	Kosrae (Micronesia)	1828	C
Aplonis fusca	Norfolk Island Starling	Norfolk I (Australia)	1925	
Aplonis mavornata	Mysterious Starling	Cook Is	1825	C/D
* *Aplonis pelzelni*	Pohnpei Mountain Starling	Pohnpei (Micronesia)	1956	B
Fregilupus varius	Réunion Starling	Réunion	1850-1860	B,C/D
Necrospar rodericanus	Rodrigues Starling	Rodrigues (Mauritius)	1726	
Family Callaeidae				
Heteralocha acutirostris	Huia	New Zealand	1907	A,B,C/D
MAMMALS				
Order MARSUPIALIA				
Family Macropodidae				
* *Caloprymnus campestris*	Desert Rat-kangaroo	Australia	1935	A,B,C
+ *Lagorchestes asomatus*	Central Hare-wallaby	Australia	1931	
Lagorchestes leporides	Eastern Hare-wallaby	Australia	1890	
Macropus greyi	Toolache Wallaby	Australia	1927	C
Onychogalea lunata	Crescent Nailtail Wallaby	Australia	1964	C,D
Potorous platyops	Broad-faced Potoroo	Australia	1875	C
Family Peramelidae				
Chaeropus ecaudatus	Pig-footed Bandicoot	Australia	1907	C,D
Perameles eremiana	Desert Bandicoot	Australia	1935	
Family Thylacomyidae				
Macrotis leucura	Lesser Bilby	Australia	1931	A,C
Family Thylacinidae				
Thylacinus cynocephalus	Thylacine	Tasmania (Australia)	1934	E
Order CHIROPTERA				
Family Pteropodidae				
Acerodon lucifer	Panay Giant Fruit Bat	Philippines	1888	
Dobsonia chapmani	Chapman's Bare-backed Flying Fox	Philippines	1964	
Pteropus pilosus	Palau Flying Fox	Palau	19th C.	
Pteropus subniger	Lesser Mascarene Flying Fox	Mauritius, Réunion		
Pteropus tokudae	Guam Flying Fox	Guam	1968	
Family Molossidae				
Mystacina robusta	NZ Lesser Short-tailed Bat	New Zealand	1960s	
Order INSECTIVORA				
Family Nesophontidae				
# *Nesophontes hypomicrus*	Atalaye Nesophontes	Haiti, Dominican Republic		C
# *Nesophontes micrus*	Western Cuban Nesophontes	Cuba		C
# *Nesophontes paramicrus*	St Michel Nesophontes	Haiti, Dominican Republic		C
# *Nesophontes zamicrus*	Haitian Nesophontes	Haiti, Dominican Republic		C
# *Nesophontes* sp.		Cayman Is		
Order LAGOMORPHA				
Family Ochotonidae				
Prolagus sardus	Sardinian Pika	Corsica, Sardinia	18th C.	

SPECIES	ENGLISH NAME	DISTRIBUTION	LAST RECORDED	POSSIBLE CAUSE
MAMMALS (Continued)				
Family Leporidae				
* Sylvilagus insonus	Omilteme Cottontail	Mexico		
Order RODENTIA				
Family Arvicolidae				
Pitymys bavaricus	Bavarian Pine Vole	Germany		
Family Capromyidae				
# Capromys sp.		Cayman Is		
# Geocapromys colombianus		Cuba		
Geocapromys thoractus		Little Swan I (Honduras)	1950s	
# Geocapromys sp.		Cayman Is		
# Isolobodon portoricensis		Haiti, Dominican Republic		
# Plagiodontia velozi		Haiti, Dominican Republic		
Family Cricetidae				
Megalomys desmarestii	Martinique Rice Rat	Martinique	1902	
Megalomys luciae	St Lucia Rice Rat	Saint Lucia	19th C.	
Megaloryzomys curioi		Galapagos (Ecuador)		
Megaloryzomys sp.		Galapagos (Ecuador)		
Nesoryzomys darwini	Santa Cruz Rice Rat	Galapagos (Ecuador)		
Nesoryzomys sp.		Galapagos (Ecuador)		
Oryzomys victus	St Vincent Rice Rat	Saint Vincent	1897	
* Peromyscus pembertoni	Pemberton's Deer Mouse	Mexico		
Family Echimyidae				
# Boromys offella		Cuba		
# Boromys torrei		Cuba		
# Brotomys voratus		Haiti, Dominican Republic		
Family Muridae				
Conilurus albipes	Rabbit-eared Tree-rat	Australia	1875	
* Crateromys paulus	Ilin Bushy-tailed Cloud-rat	Philippines		
Leporillus apicalis	Lesser Stick-nest Rat	Australia	1933	
* Notomys amplus	Short-tailed Hopping-mouse	Australia	1894	
* Notomys longicaudatus	Long-tailed Hopping-mouse	Australia	1901	
+ Notomys macrotis	Big-eared Hopping-mouse	Australia	pre-1850	
+ Notomys mordax	Darling Downs Hopping-mouse	Australia	pre-1846	
+ Pseudomys fieldi	Alice Springs Mouse	Australia	1895	
+ Pseudomys gouldi	Gould's Mouse	Australia	1930	
Rattus macleari	Maclear's Rat	Christmas I (Australia)	1908	
Rattus nativitatis	Bulldog Rat	Christmas I (Australia)	1908	
Order CARNIVORA				
Family Canidae				
Dusicyon australis	Falkland Island Wolf	Falklands Is	1876	E
Family Procyonidae				
+ Procyon gloveralleni	Barbados Racoon	Barbados		
Order PINNIPEDIA				
Family Phocidae				
Monachus tropicalis	Caribbean Monk Seal	Caribbean	1962	A
Order SIRENIA				
Family Dugongidae				
Hydrodamalis gigas	Steller's Sea Cow	Bering Straits (Russia)	1768	A
Order PERISSODACTYLA				
Family Equidae				
Equus quagga	Quagga	South Africa	1883	A,E
Order ARTIODACTYLA				
Family Bovidae				
Gazella rufina	Red Gazelle	Algeria?	19th C.	A
Hippotragus leucophaeus	Bluebuck	South Africa	1800	E
Family Cervidae				
Cervus schomburgki	Schomburgk's Deer	Thailand	1932	A

KEY:

    * Indicates species generally regarded as extinct but for which there may still be some chance of survival.

    + Indicates taxa which may be conspecific with extant forms.

    # Indicates species known from post-Columbian (i.e. post-1500) deposits in the Caribbean; some may have become extinct before 1600.

    • Indicates species last recorded from Rapa in 1934, and which were considered likely to become rapidly extinct.

    x Indicates species recorded from subfossil deposits in the Mascarenes which are considered very likely to have become extinct following settlement in 1723 although may possibly have become extinct earlier.

POSSIBLE CAUSES COLUMN:

    A   Hunting (includes for food, skin, sport, live trade, feathers)

    B   Direct habitat alteration by man

    C   Introduced predators (e.g. cats, rats, mustelids, mongooses, snails, monkeys)

    C/D Predators or others not specified

    D   Other introduced animals (e.g. goats, rabbit, pigs)

    E   Destroyed as a pest species

    F   Introduced disease

    G   Indirect effects

    H   Natural Causes; causes uncertain.

SOURCE: Compiled from multiple sources; details available from WCMC. Most bird data compiled by A Stattersfield, and kindly made available by the International Council for Bird Preservation. Mollusk data assembled by Sue Wells with the assistance of members of the SSC Mollusk Specialist Group and other malacologists.

ESSENTIAL MI-
CROBES, FUNGI, ANI-
MALS, AND PLANTS

Scientists estimate that humans utilize over 40,000 species every day. The services these species render range from food to medicine to the cycling of essential elements through the Biosphere. What follows is a list of 400 of these species (1% of the total) representing a mere sampling—a smattering—of those organisms on which we depend for our survival.

MICROBES	
Abacystis sp. cyanobacteria	Photosynthesis (oxygen/carbon cycle)
Achnanthes sp. diatom	Silica cycle
Acidaminococcus sp. fermenting bacteria	Fermentation
Actinomadura cremea actinobacteria	Medicinal: antibiotics
Amphipleura sp. diatom	Silica cycle
Amycolata autotropica actinobacteria	Medicinal: antibiotics
Anabaena sp. cyanobacteria	Photosynthesis (oxygen/carbon cycle)
Arachnoidiscus sp. diatom	Silica cycle
Asterionella sp. diatom	Silica cycle
Azomonas sp. azotobacteria	Nitrogen fixation
Azotobacter sp. azotobacteria	Nitrogen fixation
Bacillus brevis aeroendospore	Medicinal: antibiotics
Bacteroides sp. fermenting bacteria	Fermentation
Bdellovibrio sp. pseudomonad	Carbon cycle
Beijerinckia sp. azotobacteria	Nitrogen fixation
Biddulphia sp. diatom	Silica cycle
Calcidiscus sp. haptophyte	Carbon cycle
Calciosolenia sp. haptophyte	Carbon cycle
Calyptrosphaera sp. haptophyte	Carbon cycle
Chamaeliphon sp. cyanobacteria	Photosynthesis (oxygen/carbon cycle)
Chloroblum sp. anaerobic phototroph	Photosynthesis (oxygen/carbon cycle)
Chloroflexus sp. anaerobic phototroph	Photosynthesis (oxygen/carbon cycle)
Chloropseudomonas sp. anaerobic phototroph	Photosynthesis (oxygen/carbon cycle)
Chromatium sp. anaerobic phototroph	Photosynthesis (oxygen/carbon cycle)
Chroococcus sp. cyanobacteria	Photosynthesis (oxygen/carbon cycle)
Chrysochromulina sp. haptophyte	Carbon cycle
Clevelandina sp. spirochete	Cellulose metabolis
Clostridium sp. fermenting bacteria	Fermentation
Coccolithus sp. haptophyte	Carbon cycle
Coscinodiscus sp. diatom	Silica cycle
Cyclotella sp. diatom	Silica cycle
Cymbella sp. diatom	Silica cycle
Dermocarpa sp. cyanobacteria	Photosynthesis (oxygen/carbon cycle)
Derxia sp. azotobacteria	Nitrogen fixation
Desulfotomaculum sp. thiopneute	Sulfur cycle

Desulfovibrio sp. thiopneute	Sulfur cycle
Desulfuromonas sp. thiopneute	Sulfur cycle
Diatoma sp. diatom	Silica cycle
Diplocalx sp. spirochete	Cellulose metabolism
Diplococcus sp. fermenting bacteria	Fermentation
Diploneis sp. diatom	Silica cycle
Discosphaera sp. haptophyte	Carbon cycle
Emiliania sp. haptophyte	Carbon cycle
Entophysalis sp. cyanobacteria	Photosynthesis (oxygen/carbon cycle)
Eubacterium sp. fermenting bacteria	Fermentation
Eunotia sp. diatom	Silica cycle
Fragilaria sp. diatom	Silica cycle
Fusobacterium sp. fermenting bacteria	Fermentation
Gephyrocapsa sp. haptophyte	Carbon cycle
Gloeocapsa sp. cyanobacteria	Photosynthesis (oxygen/carbon cycle)
Hollandina sp. spirochete	Cellulose metabolism
Hydrogenomonas pseudomonad	Carbon cycle
Hyella sp. cyanobacteria	Photosynthesis (oxygen/carbon cycle)
Hymenomonas sp. haptophyte	Carbon cycle
Lactobacillus sp. fermenting bacteria	Fermentation
Leptotrichia sp. fermenting bacteria	Fermentation
Leuconostoc sp. fermenting bacteria	Fermentation
Lyngbya sp. cyanobacteria	Photosynthesis (oxygen/carbon cycle)
Megasphaera sp. fermenting bacteria	Fermentation
Methanosarcina sp. methanogen	Methane cycle
Microcystis sp. cyanobacteria	Photosynthesis (oxygen/carbon cycle)
Nostoc sp. cyanobacteria	Photosynthesis (oxygen/carbon cycle)
Oscillatoria sp. cyanobacteria	Photosynthesis (oxygen/carbon cycle)
Pelodictyon sp. anaerobic phototroph	Photosynthesis (oxygen/carbon cycle)
Peptococcus sp. fermenting bacteria	Fermentation
Peptostreptococcus sp. fermenting bacteria	Fermentation
Phaeocystis sp. haptophyte	Carbon cycle
Phochloron sp. chloroxybacteria	Photosynthesis (oxygen/carbon cycle)
Pillotina sp. spirochete	Cellulose metabolism
Pontosphaera sp. haptophyte	Carbon cycle
Prochlorothix sp. chloroxybacteria	Photosynthesis (oxygen/carbon cycle)
Prymnesium sp. haptophyte	Carbon cycle
Pseudomonas pseudomonad	Carbon cycle
Rhabdosphaera sp. haptophyte	Carbon cycle
Rhizobium sp. azotobacteria	Nitrogen fixation
Rhodomicrobium sp. anaerobic phototroph	Photosynthesis (oxygen/carbon cycle)
Rhodopseudomonas sp. anaerobic phototroph	Photosynthesis (oxygen/carbon cycle)
Rhodospirillum sp. anaerobic phototroph	Photosynthesis (oxygen/carbon cycle)
Ruminobacter sp. fermenting bacteria	Fermentation
Ruminococcus sp. fermenting bacteria	Fermentation
Spirulina sp. cyanobacteria	Photosynthesis (oxygen/carbon cycle)
Streptococcus sp. fermenting bacteria	Fermentation; Medicinal: antibiotics
Synechococcus sp. cyanobacteria	Photosynthesis (oxygen/carbon cycle)
Synechocystis sp. cyanobacteria	Photosynthesis (oxygen/carbon cycle)
Syracosphaera sp. haptophyte	Carbon cycle
Thiocapsa sp. anaerobic phototroph	Photosynthesis (oxygen/carbon cycle)
Thiocystis sp. anaerobic phototroph	Photosynthesis (oxygen/carbon cycle)
Thiodictyon sp. anaerobic phototroph	Photosynthesis (oxygen/carbon cycle)
Thiopedia sp. anaerobic phototroph	Photosynthesis (oxygen/carbon cycle)
Thiosarcina sp. anaerobic phototroph	Photosynthesis (oxygen/carbon cycle)

Thiothece sp. anaerobic phototroph — Photosynthesis (oxygen/carbon cycle)
Veillonella sp. fermenting bacteria — Fermentation
Xanthomonas pseudomonad — Carbon cycle
Zoogloea pseudomonad — Carbon cycle

FUNGI

Ahnfeltia plicata (alga)	Agar
Anaebaena-Azolla (alga) green manure	Fertilizer
Ascophyllum nodosum (alga) bladder wrack	Thickening agent
Asperigillus caronarius deuteromycote	Medicinal: antibiotics
Auricularia auricula brown ear fungus	Food
Calcareous algae (alga)	Fertilizer
Cantharellus cibarius chanterelle	Food
Chlamydomonas sp. (alga)	Fertilizer
Clavaria purpurea purple coral	Food
Clavariadelphus truncatus flat-top coral	Food
Claviceps purpurea ergot	Medicinal: antihemorrhagic
Codium geppii (alga) sagati	Food
Craterellus cornucopioides horn of plenty	Food
Curyillaea sp. (alga)	Thickening agent
Disciotis venosa cup morel	Food
Genea hispidula mycorrhiza structure	Nitrogen fixation
Gigartina mamillosa (alga) Irish moss	Carrageenan
Gomphus clavatus pig's ears	Food
Gracilaria sp. (alga) Lumicevata	Food
Laminaria cloustoni (alga) oar weed	Colloid stabilizer
Laminaria digitata (alga)	Thickening agent
Leotia lubrica slippery cap	Food
Macrocystis pyrifera (alga) kelp	Colloid stabilizer
Morchella conica black morel	Food
Morchella crassipes thick-footed morel	Food
Morechella esculenta common morel	Food
Pencillium chrysogenum (and other *Penicillium* sp.) penicillin molds	Medicinal: antibiotic
Phlogiotis helvelloides apricot jelly	Food
Polyozellus muliplex blue chanterelle	Food
Porphyra sp. (alga) laver	Food
Pseudohydnum gelatinosum toothjelly	Food
Psilocybe cubensis bluestain smoothcap	Hallucinogenic
Saccharomyces cerevisiae yeast	Alcohol fermentation
Sparassis crispa ruffles	Food
Spirulina sp. (alga)	Food
Tolypodcladium sp. mold	Medicinal: immunosuppressant
Tremella foliacea leaf jelly	Food
Ulva sp. (alga)	Food
Undaria pinnatifida (alga) wakame	Food
Ustilago maydis cornsmut	Food
Verpa bohemica early morel	Food
Verpa conicda bell morel	Food

ANIMALS

Acipenser oxyritynchus Atlantic sturgeon	Food
Aequipecten irradians bay scallop	Food
Albula vulpes bone fish	Food
Alces alces elk/moose	Food; Pelt; Draught

Alectis ciliaris African pompano	Food
Alectoris chuka chukar partridge	Food
Alectoris rufa red-legged partridge	Food
Alligator mississippiensis American alligator	Skin; Food
Alopex lagopus arctic fox	Pelt
Alosa sapidissima American shad	Food
Anas platyrhyunchos mallard duck	Food
Anchoa sp. anchovies	Food
Anser anser goose	Food
Anser cygnoides Chinese goose	Food
Antheraea pernyl (and other Antheraea, Attacus, and Anaphe spp.) silkworm	Industry: textile
Apis mellifera (and other Apis spp.) honeybee	Food
Arctogadus glacialis polar cod	Food
Bison bison bison	Food
Blatta orientalis cockroach	Food
Bombyx mori silkworm	Industry: textile
Bos grunniens yak	Food; Pelt
Bos indicus cattle (zebu)	Food; Pelt
Bos javanicus cattle (Javanese)	Food; Pelt
Bos taurus cattle (taurine)	Food; Pelt
Brama brama Atlantic pomfret	Food
Brevoortia sp. menhadens	Fish meal and oil
Bubalus bubalis water buffalo	Draught
Cairina moschata Muscovy duck	Food
Camelus bactrianus bactrian camel	Food; Transportation
Camelus dromedarius Dromedary	Food; Transportation
Canis familaris dog	Food; Companionship
Capra hircus goat	Food; Pelt
Cavia porcellus Guinea pig	Food; Medicinal research
Chelonia mydas green turtle	Food; Shell, Oil; Skin
Chinchilla laniger chinchilla	Pelt
Clupea harengus Atlantic herring	Food
Colinus virginianus northern quail	Food; Feathers
Columba livia pigeon	Food
Crassostrea virginca common oyster	Food
Cricetomys gambianus giant rat	Weed control
Crocodylus acutus American crocodile	Skin; Food
Crocodylus johynsoni Australian freshwater crocodile	Skin; Food
Crocodylus niloticus Nile crocodile	Skin; Food
Deladenus siricidicola (nematode worm)	Pest control: wood wasp
Dromaius novaehollandiae emu	Food; Feathers
Elephas maximus Asian elephant	Transportation
Elops saurus lady fish	Food
Equus asinus donkey	Tranportation
Equus caballus horse	Transportation; Draught
Eretmochelys imbricata hawksbill turtle	Shell
Felis catus cat	Companionship; Pest Control: mice and rats
Gadus morhua Atlantic cod	Food
Gallus gallus chicken	Food
Gerres cinereus yellowfin mojarra	Food

Heterorhabditis bacteriophora (nematode worm)	Pest control: click beetles
Heterotylenchus autumnalis (nematode worm)	Pest control: face fly
Homarus americanus norther lobster	Food
Iguana iguana green iguana	Food
Lama glama llama	Food; Pelt; Draught
Lama pacos alpaca	Food; Pelt
Loligo pialei long-finned squid	Food
Lopholatilus chamaeleonticeps tilefish	Food
Lutjanus sp. snappers	Food
Macrotermes sp. macrotermes termite	Nitrogen recycling; Celulose metabolism; Food
Mallotus villosus capelin	Food
Martes zibellina sable	Pelt
Melanogrammus aeglefinus haddock	Food
Meleagris gallopavo turkey	Food
Merluccius billinearis silver hake	Food
Mustela vison mink	Pelt
Mya arenaria soft-shelled clam	Food
Neoaplectana glaseri (nematode worm)	Pest control: Japanese beetle
Oryctolegus cuniculus: rabbit	Food; Pelt; Medicinal research
Osmerus mordax rainbow smelt	Food
Ovibos moschatus musk ox	Food; Pelt
Ovis aries sheep	Food; Pelt
Penaeus duorarum pink shrimp	Food
Perdix perdix common partridge	Food; Feathers
Phasianus colchicus common pheasant	Food; Feathers
Pomatomus saltatrix bluefish	Food
Pristionchus uniformis (nematode worm)	Pest control: Colorado beetle
Rana spp. frogs	Food
Rangifer tarandus reindeer	Food; Transportation; Pelt
Romanomermis culicivorax (nematode worm)	Pest control: mosquitoes
Salmo gairdreri rainbow trout	Food
Salmo salar Atlantic salmon	Food
Salmo trutta brown trout	Food
Salvelinus alpinus arctic char	Food
Salvelinus fontinalis brook trout	Food
Sardinella sp. sardines	Food
Struthio camelus ostrich	Food; Feathers
Sus domesticus pig	Food
Sus scrofa wild boar	Food
Tripius sciarae (nematode worm)	Pest control: sciarid flies
Vicugna vicugna vicuna	Pelt
Vulpes vulpes red/silver fox	Pelt

PLANTS

(All plants photosynthesize and are therefore major players in the oxygen/carbon cycle.)

Adonia vernalis pheasant's eye	Medicinal: cardiotonic
Aesculus hippocastanum horse chestnut	Medicinal: antiinflammatory
Agrimonia eupatoria common agrimony	Medicinal: anthelmintic
Aloe sp. aloe	Medicinal: antiinflammatory
Allium cepa onion	Food

Allium fistulosum onion	Food
Allium sativum garlic	Food
Ammi visnaga toothpick plant	Medicinal: bronchodilator
Anabasis aphylla tumbleweed	Medicinal: skeletal muscle relaxant
Ananas comosus pineapple	Food; Medicinal: antiinflammatory
Andrographis paniculata karyat	Medicinal: antibacterial
Anisodus tanguticus zàng qìè	Medicinal: anticholinergic
Arachis hypogaea groundnut	Food
Areca catechu betel-nut palm	Medicinal: anthelmintic
Atropa belladonna belladonna	Medicinial: anticholinergic
Avena sativa oats	Food
Berberis vulgaris barberry	Medicinal: antibacterial
Bertholetia excelsa Brazil nut	Food
Beta vulgaris sugar beet	Food
Bibes rubrum currants	Food
Brassica juncea mustard seed	Food
Brassica nigra black mustard	Medicinal: rubefacient
Brassica oleracea cabbage	Food
Brassica rapa cabbage	Food
Cajanus cajan pigeonpea	Food
Camellia sinensis tea	Food; Medicinal: stimulant
Cannabis sativa hemp	Medicinal: antiemetic
Capsicum annuum chili pepper	Food
Caroca papaya papaya	Medicinal: proteolytic
Carthamus tinctorius safflowerseed	Food
Cassia spp. senna	Medicinal: laxative
Centella asiatica Indian pennywort	Medicinal: vulnerary
Cephaelis ipecacuanha ipecac	Medicinal: amebicide
Chenopodium quinoa quinos	Food
Cicer arietinum chickpea	Food
Cinchona sp. quinine	Medicinal: antimalarial
Cinnamomum camphora camphor tree	Medicinal: rubefacient
Citrkus reticulata tangerine	Food
Citrus aurantifolia lime	Food
Citrus grandis pomelo	Food
Citrus limon lemon	Food; Medicinal: antihemorrhagic
Citrus paradisi grapefruit	Food
Citrus sinensis orange	Food; Medicinal: antihemorrhagic
Cocos nucifera coconut	Food
Coffea arabica coffee	Food
Colchicum autumnale autumn crocus	Medicinal: antitumor agent
Colocasia esculenta taro	Food
Convallaria majaliks lily-of-the-valley	Medicinal: cardiotonic
Coptis japonica goldthread	Medicinal: antipyretic
Corylus avellana Hazel	Food
Corylus maxima filbert	Food
Cucumis sativus cucumber	Food
Cucumis sativus watermelon	Food
Cucurbita maxima pumpkin	Food
Cucurbita moschata squash	Food
Cucurbita pepo gourd	Food
Curcuma longa turmeric	Medicinal: choleretic
Cynara scolymus artichoke	Food
Daucus carota carrot	Food
Digitalis purprea foxglove	Medicinal: cardiotonic

Digitaria exilis fonio	Food
Dioscorea spp. yam	Food
Diospyros spp. ebony	Industry: timber
Duboisia sp. corkwood	Medicinal: antiinflammatory
Echinochloa frumentacea Japanese barnyard millet	Food
Elaeis guineensis oil palm	Food
Elettaria cardamomum cardamon	Food
Eleusine coracana finger millet	Food
Ephedra sinica ma-huang	Medicinal: sympathomimetic
Erythroxylum coca coca	Medicinal: anesthetic
Ficus carica fig	Food
Fragaria ananassa strawberry	Food
Fraxinus rhynchophylla	Medicinal: antidysentery
Gaultheria procumbens wintergreen	Medicinal: rubefacient
Glaucium flavum horned poppy	Medicinal: antitussive
Glycine max soybean	Food
Glycyrrhiza blabra licorice	Food
Gossypium sp. cotton	Industry: texile; Medicinal: male contraceptive
Gossyplum barbadense cottonseed	Food
Helianthus annuus sunflower seed	Food
Hemsleya amabilis luó guo di	Medicinal: antibacterial
Hordeum vulgare barley	Food
Hydrangea macrophylla hydrangea	Food: sweetener
Hydrastis canadensis golden seal	Medicinal: hemostatic
Hyoscyamus niger henbane	Medicinal: anticholinergic
Ilex paraguariensis mate	Food
Illicium verum star anise	Food
Ipomoea batates sweet potato	Food
Juglans regia walnut	Food
Juniperus oxycedrus cedar	Medicinal: anesthetic
Lablab purpureus lablab bean	Food
Lactuca sativa lettuce	Food
Larrea divaricasta creosote bush	Medicinal: antioxidant
Lens culinaris lentil	Food
Lobelia inflata Indian tobacco	Medicinal: respiratory stimulant
Lubinus mutabilis lupin	Food
Lycopersicon esculentum tomato	Food
Lycoris squamigera resurrection lily	Medicinal: cholinesterase inhibitor
Malus pumilla apple	Food
Manfigera indica mango	Food
Manihot esculenta cassava	Food
Medicago sativa alfalfa	Fodder
Mentha spp. mint, e.g., peppermint, spearmint	Medicinal: rubefacient; Food: flavoring, tea
Mucuna deeringiana velvet bean	Medicinal: antiparkinsonism
Musa acuminate banana	Food
Musa paradisaca plantain	Food
Myroxylon balsamum balsam	Medicinal: diuretic
Ochroma pyramidale balsa	Industry: timber
Ocotea glaziovii yellow cinnamon	Medicinal: antidepressant
Olea europaea olive	Food
Oryza glaberrma rice	Food
Oryza sativa rice	Food

Panicum millaceum common millet	Food
Papaver somniferum opium poppy	Medicinal: analgesic
Pennisetum americanum bulrush millet	Food
Persea americana avocado	Food
Phaseolus lunatus lima bean	Food
Phaseolus vulgaris haricot bean	Food
Phoenix dactylifera date	Food
Physostigma venenosum ordeal bean	Medicinal: anticholinesterase
Pimenta dioica pimento	Food
Pinus elliotti slash pine	Industry: timber
Pinus radiat Monterrey pine	Industry: timber
Pinus strobus white pine	Industry: timber
Pinus taeda loblolly pine	Industry: timber
Piper methysticum kava	Medicinal: tranquilizer
Piper nigrum pepper	Food
Pistecia vera pistachio	Food
Pisum sativum pea	Food
Plantago sp. plantago	Medicinal: laxative
Podophyllum peltatum May apple	Medicinal: antitumor agent
Potentilla fragariodes cinquefoil	Medicinal: hemostatic
Prunus amygdalus almond	Food
Prunus armeniaca apricot	Food
Prunus avium cherry	Food
Prunus communis pear	Food
Prunus domestica plum	Food
Prunus persica peach	Food
Pseudotsuga menziesu douglas fir	Industry: timber
Rauvokfia serpentina Indian snakeroot	Medicinal: circulatory stimulant
Rauvolfia tetraphylla snakeroot	Medicinal: antihypertensive
Ribes nigrum currants	Food
Ricinus communis castor bean	Medicinal: laxative
Saccarhum officinarum sugarcane	Food
Secale cereale rye	Food
Sesamum orientale sesame seed	Food
Setaria italica foxtail millet	Food
Simaruba glauca paradise tree	Medicinal: amebicide
Solanum melongena eggplant	Food
Solanum tuberosum pototo	Food
Sophora pachycarpa pagoda tree	Medicinal: oxytocic
Sorghum bicolor sorghum	Food
Spinacia olerecia spinach	Food
Strophanthus gratus twisted flower	Medicinal: cardiotonic
Styrax sp. benzoin	Medicinal: antiseptic
Tectona grandis teak	Industry: timber
Theobroma cacao cocoa	Food
Thymus vulgaris thyme	Food: flavoring; Medicinal: expectorant
Triticum aestivum wheat	Food
Triticum turgidum wheat	Food
Veratrum sp. hellebore	Medicinal: relaxant
Vicia faba broad bean	Food
Vigna unguiculata cowpea	Food
Vitellaria paradoxa karite nut	Food
Vitis vinifera grape	Food
Xanthostoma sagittifolium Tautia	Food
Zea mays maize (corn)	Food

BIBLIOGRAPHY
OF SUGGESTED
READINGS

GENERAL BOOKS ON BIODIVERSITY

A number of titles have appeared in recent years dealing with the four big questions: What Is Biodiversity? What are its values—or Why should we care about biodiversity? What is causing the biodiversity crisis—the Sixth Extinction? And, What can we do about it? The following are among the more prominent general contributions:

Eldredge, N. 1995. *Dominion.* Henry Holt, New York. Reprinted 1997, University of California Press, Berkeley. Here I present in detail the story of human ecological evolution, the rise of culture as the dominant factor in the human ecological "niche"—and the impact of agriculture on human life, population growth, and the rest of the planet.

Reaka-Kudla, M. L., Wilson, D. E., and Wilson, E. O. 1996. *Biodiversity II. Understanding and Protecting Our Biological Resources.* Joseph Henry Press, Washington, D.C.

Wilson, E. O. 1993. *The Diversity of Life.* Harvard University Press, Cambridge, Massachusetts. Of all of the recent books on biodiversity, perhaps the most significant so far—for ease of reading and coherent insight—is this book by the acknowledged guru of biodiversity.

Chapter 1: Tales from the Swamp
The Kalahari and Okavango Systems

Campbell, A. 1990. *The Nature of Botswana: A Guide to Conservation and Development.* International Union for Conservation of Nature and Natural Resources (IUCN) Field Operations Division, Gland, Switzerland. A balanced view of humanity and nature in today's Botswana—written by a truly wise man who knows and cares.

Main, M. 1987. *Kalahari: Life's Variety in Dune and Delta.* Southern Book Publishers, Johannesburg. By far the best general discussion of the Kalahari and Okavango, their climate, plants and animals, their people—and the problems facing both the Kalahari and Okavango ecosystems.

Scholz, C. 1997. *Fieldwork: A Geologist's Memoir of the Kalahari.* Princeton University Press, Princeton. Adventures of a young seismologist in Botswana whose work confirmed the notion that the Okavango Delta lies in the southwesternmost extension of the Rift Valley System.

The Human Evolutionary Story

Johanson, D. and Blake, E. 1996. *From Lucy to Language.* Nevraumont/Simon and Schuster, New York. A beautifully illustrated account of hominid evolution.

Tattersall, I. 1995. *The Fossil Trail: How We Know What We Think We Know About Human Evolution.* Oxford University Press, New York and Oxford. An engaging account of discovery of the human fossil record—and of the endless search to learn how to understand our own history.

Tattersall, I. 1998. *Becoming Human: Evolution and Human Uniqueness.* Harcourt Brace and Co., New York. A brilliant analysis of how, why and when uniquely human traits—especially intelligence and creativity—appeared in human evolutionary history.

Chapter 2: Biodiversity, Evolution, and Ecology

No single book addresses evolution and ecology, and their interrelationships, in the manner of this chapter—though my own Reinventing Darwin *comes close to filling this niche.*

Eldredge, N. 1992 (editor). *Systematics, Ecology and the Biodiversity Crisis.* Columbia University Press, New York. A technical symposium evaluating, through some 13 separate contributions, the relationship between ecology and evolution—and how both, severally and together, pertain to understanding and ameliorating the biodiversity crisis.

Eldredge, N. 1995. *Reinventing Darwin.* John Wiley and Sons, New York. My personal account of modern evolutionary biology—including conflicts between paleontologists (such as myself) and geneticists.

Futuyma, D. J. 1997. *Evolutionary Biology.* Sinauer Associates, Sunderland, Massachusetts. An up-to-date college text on evolution.

Odum, E. 1983. *Basic Ecology.* Third edition, Saunders College Publishing, Philadelphia. A classic text on ecology.

Chapter 3: The Tree of Life

By far the best overview of the evolutionary diversity of life is the third edition of The Five Kingdoms—*originally authored solely by biologist Lynn Margulis.*

Margulis, L. and Schwartz, K. V. 1998. *The Five Kingdoms.* Third edition. W. H. Freeman, New York. An authoritative review of all the major phyla of life with extensive bibliographies.

The following titles focus on different major segments of the evolutionary diversity of life:

Brock, T. D. and Madigan, M. 1990. *The Biology of Microorganisms.* Sixth edition, Prentice-Hall, Paramus, New Jersey.

Brusca, Richard C. and Brusca, Gary J. 1990. *The Invertebrates.* Sinauer Associates, Sunderland, Massachusetts.

Hausmann, K. and Hulsmann, N. 1995. *Protozoology.* Thieme Medical Publications, New York.

Margulis, L. and Sagan, D. 1986. *Microcosmos.* Summit Books, New York.

Raven, P. 1998. *Biology of Plants.* Fifth edition, Worth Publications, New York.

Young, J. Z. 1991. *The Life of Vertebrates.* Third edition, Oxford University Press, Oxford.

Chapter 4: Ecosystem Panorama

Instead of listing separate books on each of the major types of ecosystems discussed in this chapter, I will cite an excellent 30-volume series that does the job in stunning detail. I also include two general works on the global ecosystem.

Goodall, D. W., Editor-in-Chief. *Ecosystems of the World.* Elsevier, Amsterdam. This comprehensive series, still in production, provides in-depth coverage of all the major types of ecosystems on earth.

Lovelock, J. 1979. *Gaia.* Oxford University Press, Oxford. A forceful and imaginative presentation of the entire globe—atmosphere, hydrosphere, lithosphere and biosphere—integrated into a single, complex system.

Vernadsky, V. 1998. *The Biosphere.* Nevraumont/Copernicus, Springer-Verlag, New York. A new edition of a classic work in which the concept of the Biosphere as it is understood today was first enunciated.

Chapter 5: Biodiversity—A Threatened Natural Treasure

The values of—and threats to—biodiversity are covered at length in a number of the books already cited under general biodiversity above. In addition, I recommend:

Alvarez, W. 1997. T. rex *and the Crater of Doom.* Princeton University Press, Princeton. An engaging account of the mass extinction 65 million years ago—and the scientific detective work leading to the solution to the puzzle: What killed the dinosaurs?

Eldredge, N. 1991. *The Miner's Canary: Unraveling the Mysteries of Extinction.* Reprinted 1994, Princeton University Press, Princeton. A survey of mass extinctions of the geological past, and their effects on the evolution of life—written to shed light on the current biodiversity crisis.

Kellert, S. R. 1996. *The Value of Life. Biological Diversity and Human Society.* Island Press, Washington, D.C. and Covelo, California. How the American public views issues of biodiversity and conservation.

Norton, B. G. 1987. *Why Preserve Natural Variety?* Princeton University Press, Princeton. A thoughtful—and thought-provoking—examination of the values of biodiversity.

Chapter 6: Striking a Balance

Once again, books such as Wilson's The Diversity of Life *and my own* Dominion *discuss ways to stem the tide of the Sixth Extinction. In addition, I suggest the following—with special attention to the timeless appeal and relevance of the two classics by Carson and Leopold.*

Carson, R. L. 1962. *Silent Spring.* Reprinted 1994, Houghton Mifflin, Boston. A stirring call to arms against the destructive impact of DDT—but also a timeless message on the human impact on the planet, and what can be done about it.

Cohen, J. E. 1995. *How Many People Can the Earth Support?* W.W. Norton, New York. A masterful review of human population—past, present and future—and what we might do to regulate our own numbers before the global system imposes its own controls on us.

Dobson, A. P. 1996. *Conservation and Biodiversity.* Scientific American Library, New York. Threats to biodiversity—and what can be done to preserve the living world.

Leopold, A. 1949. *A Sand County Almanac.* Oxford University Press, New York. A soliloquy on land and life that cuts to the very heart of the values of the natural world, and the need to preserve it.

INDEX